The Open University

The
Molecular World

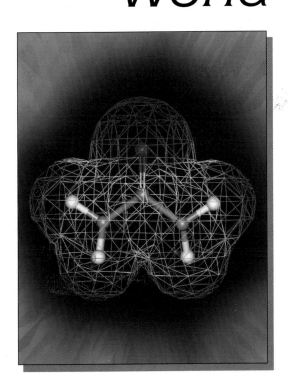

Molecular Modelling and Bonding

edited by Elaine Moore

This publication forms part of an Open University course, S205 *The Molecular World*. Most of the texts which make up this course are shown opposite. Details of this and other Open University courses can be obtained from the Call Centre, PO Box 724, The Open University, Milton Keynes MK7 6ZS, United Kingdom: tel. +44 (0)1908 653231, e-mail ces-gen@open.ac.uk

Alternatively, you may visit the Open University website at http://www.open.ac.uk where you can learn more about the wide range of courses and packs offered at all levels by The Open University.

The Open University, Walton Hall, Milton Keynes, MK7 6AA

First published 2002

Edited, designed and typeset by The Open University.

Published by the Royal Society of Chemistry, Thomas Graham House, Science Park, Milton Road, Cambridge CB4 0WF, UK.

Printed in the United Kingdom by Bath Press Colourbooks, Glasgow.

ISBN 0 85404 675 5

A catalogue record for this book is available from the British Library.

1.1

s205book 6 i1.1

The Molecular World

This series provides a broad foundation in chemistry, introducing its fundamental ideas, principles and techniques, and also demonstrating the central role of chemistry in science and the importance of a molecular approach in biology and the Earth sciences. Each title is attractively presented and illustrated in full colour.

The Molecular World aims to develop an integrated approach, with major themes and concepts in organic, inorganic and physical chemistry, set in the context of chemistry as a whole. The examples given illustrate both the application of chemistry in the natural world and its importance in industry. Case studies, written by acknowledged experts in the field, are used to show how chemistry impinges on topics of social and scientific interest, such as polymers, batteries, catalysis, liquid crystals and forensic science. Interactive multimedia CD-ROMs are included throughout, covering a range of topics such as molecular structures, reaction sequences, spectra and molecular modelling. Electronic questions facilitating revision/consolidation are also used.

The series has been devised as the course material for the Open University Course S205 *The Molecular World*. Details of this and other Open University courses can be obtained from the Course Information and Advice Centre, PO Box 724, The Open University, Milton Keynes MK7 6ZS, UK; Tel +44 (0)1908 653231; e-mail: ces-gen@open.ac.uk. Alternatively, the website at www.open.ac.uk gives more information about the wide range of courses and packs offered at all levels by The Open University.

Further information about this series is available at www.rsc.org/molecularworld.

Orders and enquiries should be sent to:

Sales and Customer Care Department, Royal Society of Chemistry, Thomas Graham House, Science Park, Milton Road, Cambridge, CB4 0WF, UK

Tel: +44 (0)1223 432360; Fax: +44 (0)1223 426017; e-mail: sales@rsc.org

The titles in *The Molecular World* series are:

THE THIRD DIMENSION
edited by Lesley Smart and Michael Gagan

METALS AND CHEMICAL CHANGE
edited by David Johnson

CHEMICAL KINETICS AND MECHANISM
edited by Michael Mortimer and Peter Taylor

MOLECULAR MODELLING AND BONDING
edited by Elaine Moore

ALKENES AND AROMATICS
edited by Peter Taylor and Michael Gagan

SEPARATION, PURIFICATION AND IDENTIFICATION
edited by Lesley Smart

ELEMENTS OF THE p BLOCK
edited by Charles Harding, David Johnson and Rob Janes

MECHANISM AND SYNTHESIS
edited by Peter Taylor

The Molecular World Course Team

Course Team Chair
Lesley Smart

Open University Authors
Eleanor Crabb (Book 8)
Michael Gagan (Book 3 and Book 7)
Charles Harding (Book 9)
Rob Janes (Book 9)
David Johnson (Book 2, Book 4 and Book 9)
Elaine Moore (Book 6)
Michael Mortimer (Book 5)
Lesley Smart (Book 1, Book 3 and Book 8)
Peter Taylor (Book 5, Book 7 and Book 10)
Judy Thomas (*Study File*)
Ruth Williams (skills, assessment questions)
*Other authors whose previous contributions to the earlier
courses S246 and S247 have been invaluable in the
preparation of this course:* Tim Allott, Alan Bassindale, Stuart
Bennett, Keith Bolton, John Coyle, John Emsley, Jim Iley, Ray
Jones, Joan Mason, Peter Morrod, Jane Nelson, Malcolm
Rose, Richard Taylor, Kiki Warr.

Course Manager
Mike Bullivant

Course Team Assistant
Debbie Gingell

Course Editors
Ian Nuttall
Bina Sharma
Dick Sharp
Peter Twomey

CD-ROM Production
Andrew Bertie
Greg Black
Matthew Brown
Philip Butcher
Chris Denham
Spencer Harben
Peter Mitton
David Palmer

BBC
Rosalind Bain
Stephen Haggard
Melanie Heath
Darren Wycherley

Tim Martin
Jessica Barrington

Course Reader
Cliff Ludman

Course Assessor
Professor Eddie Abel, University of Exeter

Audio and Audiovisual recording
Kirsten Hintner
Andrew Rix

Design
Steve Best
Carl Gibbard
Sarah Hack
Mike Levers
Sian Lewis
John Taylor
Howie Twiner

Library
Judy Thomas

Picture Researchers
Lydia Eaton
Deana Plummer

Technical Assistance
Brandon Cook
Pravin Patel

Consultant Authors
Ronald Dell (*Case Study:* Batteries and Fuel Cells)
Adrian Dobbs (Book 8 and Book 10)
Chris Falshaw (Book 10)
Andrew Galwey (*Case Study:* Acid Rain)
Guy Grant (*Case Study:* Molecular Modelling)
Alan Heaton (*Case Study:* Industrial Organic Chemistry,
    Case Study: Industrial Inorganic Chemistry)
Bob Hill (*Case Study:* Polymers and Gels)
Roger Hill (Book 10)
Anya Hunt (*Case Study:* Forensic Science)
Corrie Imrie (*Case Study:* Liquid Crystals)
Clive McKee (Book 5)
Bob Murray (*Study File*, Book 11)
Andrew Platt (*Case Study:* Forensic Science)
Ray Wallace (*Study File*, Book 11)
Craig Williams (*Case Study:* Zeolites)

CONTENTS

MOLECULAR MODELLING AND BONDING

Elaine Moore

CASE STUDY: MOLECULAR MODELLING IN RATIONAL DRUG DESIGN

Guy Grant and Elaine Moore

INTRODUCTION

There is no substitute for practical laboratory experience, but computer modelling methods play an important role both as an aid in interpreting experimental results and as a means of explaining these results. Molecular modelling is now used not just in chemistry but in a wide range of subjects such as pharmacology and mineralogy. Figure 1.1, for example, shows a computer model of the protein haemoglobin.

One example of the computer modelling approach is to provide a reasonably accurate first guess at a structure which can then be used in methods such as X-ray diffraction of powders which, unlike the X-ray diffraction of single crystals, do not provide enough information to determine the total structure from scratch.

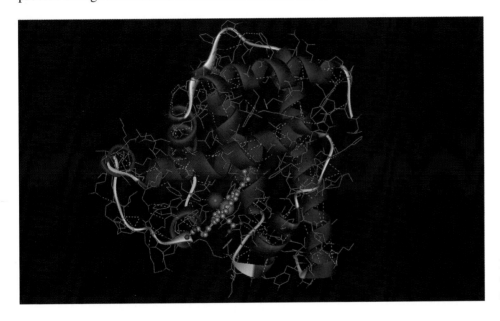

Figure 1.1
Computer model of haemoglobin, a protein. 🖥 *

In the field of drug design, computer methods have been used to screen for potentially active compounds or suggest modifications of known compounds that would be more active. Testing of inactive compounds can be avoided.

Zeolites and other structures with cages or channels will accommodate some molecules at the expense of others. Modelling of these structures has enabled planned synthesis of required molecules by predicting the cavity needed to produce the molecule.

There are areas of chemistry where it is difficult or impossible to obtain experimental results. A good example of this is the determination of the nature of reaction intermediates in chemical kinetics. In favourable cases, reaction intermediates can be studied spectroscopically using specialized techniques that allow observations on a very short time-scale of the order 10^{-9}–10^{-15} seconds.

* This symbol, 🖥, indicates that this Figure can be viewed using WebLab ViewerLite™ or Orbital Viewer from the CD-ROM associated with this Book.

However, in the majority of cases, the path of a reaction is inferred from the stereochemistry or distribution of products and the effect of changing reaction conditions such as the polarity of the solvent. Modern computational techniques allow us to follow a reaction path. In solids very simple techniques can be used to find the most favourable path for diffusion of ions. At the other extreme, very sophisticated and accurate molecular orbital techniques can calculate the energy along the entire reaction path from reactant(s) to product(s) for gas phase reactions of small molecules. More often modelling is used to calculate the relative stabilities of proposed intermediates.

It is possible to model molecules or structures that only exist at high temperatures and/or high pressures or are too dangerous to handle. Theoretical methods have been used, for example, to predict that at the temperatures and pressures in the interior of the planet Jupiter (Figure 1.2), hydrogen can exist as a liquid metal. Conditions under which such a form of hydrogen would occur were not achieved experimentally on Earth until 1996.

You are going to look in this Book at some of the methods used to model molecules and at the principles behind them. On the CD-ROM, you will study some examples.

We start by looking at a simple, but surprisingly effective, model that is based on a picture of molecules as charged spheres linked by springs.

Figure 1.2
Jupiter as viewed by the Voyager spacecraft.

MOLECULAR MECHANICS

Molecular mechanics has proved particularly useful for studying large molecules and crystalline solids where more accurate methods are very demanding of computer resources.

We shall look at the principles behind this method by considering some examples. First we consider a simple ionic solid, then move to an example where some allowance for covalency is required. Finally we consider the modelling of organic molecules.

2.1 Ionic solids

The starting point for the application of molecular mechanics to ionic solids is similar to the starting point for lattice energy calculations.* Indeed the method can be used to calculate lattice energies, but it is also used to study the effect of defects, the nature of crystal surfaces and properties of crystals.

As for lattice energies, we start by placing the ions of the crystal on their lattice sites.

● What is the force holding ions together in a crystal?

● Electrostatic attraction between the positively- and negatively-charged ions.

The ions are assumed to be on their lattice sites with their formal charges, so that in NaCl, for example, we have an array of Na^+ and Cl^- ions. The net interaction can be obtained by summing the interactions over all the pairs of ions, including not only the attraction between Na^+ and Cl^- but also the repulsion between ions of the same sign. The net interaction decreases with distance but slowly so that it is difficult to obtain an accurate value.

To calculate lattice energies, this summation can be achieved for simple lattice structures by introducing the Madelung constant. However, for layer structures with low symmetry this approach is not feasible, as a single Madelung constant will not suffice. Since the computer programs in use are set up to be of general application, they employ methods that give a good approximation to the sum over an infinite lattice for any unit cell.

However, electrostatic interaction is not all that has to be considered. We know, for example, that ions are not just point charges but have a size; the shell of electrons around each nucleus prevents too close an approach by other ions. We therefore include a term to allow for the interaction between shells on the different ions. It would be possible to give each ion a fixed size and insist that the ions cannot be closer than their combined radii. However, most programs use a different approach by including terms representing intermolecular forces.

---

* The determination of lattice energies is discussed in more detail in *Metals and Chemical Change*.[1] See the references in Further Reading (p. 123) for details of other titles in *The Molecular World* series that are relevant.

The intermolecular forces act between cations, and between cations and anions, as well as between anions. For oxides in particular, however, the cation–cation term is often ignored.

Salts such as magnesium oxide can be thought of as close-packed* arrays of anions with cations occupying the octahedral holes.

* Close packed arrays are discussed in *The Third Dimension*.[2]

Do the cations come into contact with each other in such a structure?

No, not if the anions have the larger ionic radius.

Because the cations are held apart by the anions, the cation–cation interaction is unimportant.

The final thing we need to take into account is the polarizability of the ions. This is a measure of how easily the ions are deformed from their normal spherical shape. In a perfect crystal, the ions are in very symmetrical environments and can be thought of as spherical. If one ion moves to an interstitial site, leaving its original position vacant, then the environment may not be so symmetrical and it may be deformed by the surrounding ions. A very simple way to model this is to divide the ionic charge between a core that stays fixed at the position of the ion and a surrounding shell that can move off-centre. The distribution of the charge is obtained by adjustment to fit the properties of a crystal containing that ion. The model is illustrated in Figure 2.1. The shell behaves as though it were attached to the core by springs. Take a chloride ion, for example. If the surrounding ions move so that there is a greater positive charge in one direction, then the shell will move so that the total charge on the ion is distributed over two centres producing a dipole. Opposing this will be the pull of the springs that attach it to the core.

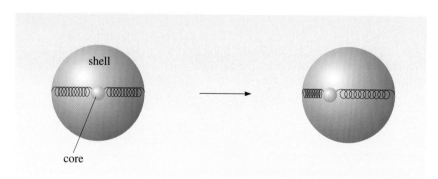

Figure 2.1
Shell model for a polarizable ion.

For ionic solids, the most important term for lattice energies is the electrostatic term; for sodium chloride, for example, the total lattice energy in a typical calculation is $-762.073 \text{ kJ mol}^{-1}$, of which $-861.135 \text{ kJ mol}^{-1}$ is due to the electrostatic interaction while the intermolecular force and shell terms contribute $+99.062 \text{ kJ mol}^{-1}$. Thus the contributions of the intermolecular force and shell terms are about 10% of the electrostatic interactions. These other terms may have a greater relevance in the study of defects.

2.1.1 Crystal defects in silver chloride

Silver halides are used in photography to capture light and form an image. The action of light on the halide produces silver which forms the black areas of the negative (Figure 2.2). The formation of silver depends on the presence of Frenkel defects in the crystal.

Figure 2.2
A black and white photograph and its negative.

- The two most common point defects in crystals are Frenkel defects and Schottky defects. What are these?

- In Schottky defects, equal numbers of cations and anions are missing (for 1 : 1 structures such as AgCl). In Frenkel defects, an ion is displaced from its lattice site to an interstitial site; for example, a small cation in a crystal with the NaCl structure can move to a tetrahedral hole from the octahedral hole normally occupied.

We can use molecular mechanics to estimate the energies of these defects in silver halides.

The dominant defect in silver halides is a Frenkel defect, in which a silver ion moves to an interstitial site. To calculate the energy required to form this defect we simply remove a silver ion from one position, put it in its new position and compare the energy of the crystal lattice with that of the perfect lattice.

- In a Frenkel defect, there is a vacancy where an ion should be and an ion in a more crowded interstitial position. Would you expect the ions in the vicinity of the defect to stay on their lattice positions?

- It would be reasonable to suppose that the ions would adjust their positions to allow the interstitial atom more room, and to take up the space left by the vacancy.

When calculating the energy of formation of the defect the nearest atoms are allowed to adjust their position to obtain the lowest energy for the crystal including the defect. Figure 2.3 (overleaf) shows how the chloride ions move when a Frenkel defect forms in AgCl; in the perfect crystal (Figure 2.3a) there is just one Ag—Cl bond length, whereas in the defect crystal (Figure 2.3b) the Ag—Cl bond lengths are shortened and variable.

For an estimate of the actual numbers of defects we need to know the Gibbs energy of formation, but the major contribution comes from the internal energy. Calculated values for the energy of formation of cation Frenkel defects in NaCl and AgCl are $308 \, \text{kJ mol}^{-1}$ and $154 \, \text{kJ mol}^{-1}$, respectively.

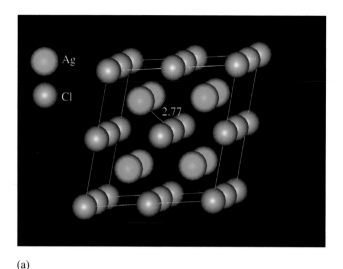

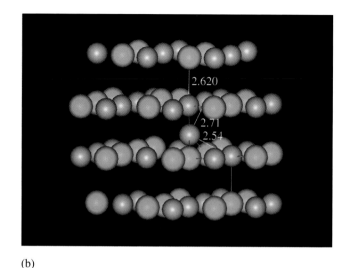

(a)   (b)

Figure 2.3 (a) Perfect AgCl. (b) A Frenkel defect in AgCl. 💻

🔘 Will NaCl or AgCl be expected to contain more Frenkel defects at room temperature?

🔘 AgCl. The defect energy is much lower.

The energy for Schottky defects in which there are cation and anion vacancies is $222\,kJ\,mol^{-1}$ for NaCl.

🔘 Will NaCl have more Schottky or more Frenkel defects?

🔘 Schottky. These are of lower energy.

QUESTION 2.1

List the types of force that would be employed in a typical molecular mechanics calculation of Schottky defects in caesium fluoride.

QUESTION 2.2

The energies of Schottky defects in KF, KCl, KBr and KI have been calculated as 244, 241, 219 and $210\,kJ\,mol^{-1}$, respectively. Which halide will have the most defects at room temperature?

STUDY NOTE

A set of interactive self-assessment questions is provided on the *Molecular Modelling and Bonding* CD-ROM. The questions are scored, and you can come back to the questions as many or as few times as you wish in order to improve your score on some or all of them, and this is a good way of reinforcing the knowledge you have gained while studying this Book.

2.1.2 Zeolites

Zeolites[*] have frameworks of silicon, aluminium and oxygen atoms which form channels and cages, e.g. Figure 2.4. They form a wide variety of structures but all are based on silicon tetrahedrally bound to oxygen. Differing numbers of silicon

* Zeolites are discussed in some detail in the Case Study in *Chemical Kinetics and Mechanism*.[3]

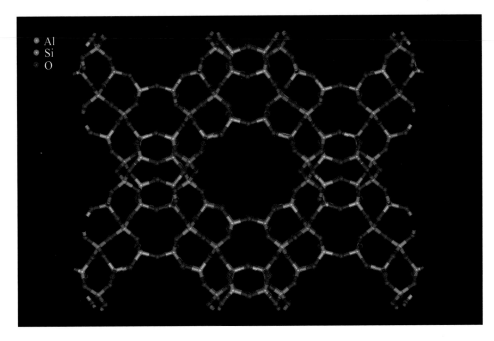

Figure 2.4
A zeolite, faujasite. Note the large channel in the centre of the structure, and the smaller ones surrounding it. Some or all of these channels can be occupied by balancing cations and by molecules. Occupation of these channels by molecules leads to the use of zeolites as catalysts. 🖳

atoms are replaced by aluminium. Other cations, notably those of Groups I, II and the lanthanides, are present in the structures to balance the charge.

Surprisingly such structures can be very successfully modelled by considering them as a collection of ions and using the methods discussed in the previous Section.

🔘 Why is this surprising?

🔘 Silicon is not normally thought of as forming Si^{4+} ions; indeed silica, SiO_2, and silicates do contain silicon covalently bonded to oxygen.

We do have to make some allowance for the covalency of the Si—O bonds. The most successful way of doing this is to add a term that represents the resistance of $\angle OSiO$ and $\angle OAlO$ bond angles to deviation from the tetrahedral angle. The covalency of zeolites and related compounds is also reflected in the relative size of the electrostatic and other terms. For one form of silica, SiO_2, for example, a calculated lattice energy of $-12\,416.977\,kJ\,mol^{-1}$ had contributions of $-16\,029.976\,kJ\,mol^{-1}$ from electrostatic interactions, $+3\,553.796\,kJ\,mol^{-1}$ from intermolecular force terms and the core–shell spring term, and $1.913\,kJ\,mol^{-1}$ from those $\angle OSiO$ bond angles that were not tetrahedral. Here the intermolecular force terms are about 20% of the electrostatic interaction. The energy due to the term keeping the angles tetrahedral is small, but without this term the zeolite structure is lost.

With this addition, the structures of a wide variety of zeolites, both naturally occurring minerals and synthetic zeolites tailored to act as catalysts, can be modelled and then used to answer questions such as which position will the non-framework ions and molecules occupy and how do ions travel through the structure?

One example of the use of molecular mechanics is in investigating the mechanism of oxygen diffusion in albite, $NaAlSi_3O_8$, a mineral related to zeolites. Experiments indicate that diffusion of oxygen in albite is faster in the presence of water. Molecular mechanics calculations show that the activation energy for OH^- diffusing through the solid is lower than that for diffusion for O^{2-}.

2.2 Modelling organic molecules

Molecular mechanics modelling of organic molecules is well-developed and widely used in the pharmaceutical industry, where it is employed to model drugs and their interactions with receptors.

The power of this method for organic molecules lies in the adoption of a relatively small set of parameters that can be transferred to any molecule you want.

But what sort of parameters might be needed? Can we simply use electrostatic and intermolecular forces? How do we allow for bonds and different conformations?

Let us start by looking at a very simple molecule — ethane. Ethane is H_3C-CH_3. As for solids, we do need to include an electrostatic interaction, but what charge are we going to give carbon and hydrogen atoms? Obviously +4 or −4 on C and +1 or −1 on H are unrealistic and would not even give a neutral molecule. Think for a moment about the process of bond formation. When two atoms form a covalent bond, they share electrons. If the atoms are unlike then one atom has a larger share than the other, resulting in a positive charge on one atom and a negative charge on the other. But the charge transferred is less than one electron. For diatomic molecules, the charge on each atom can be obtained experimentally. In the molecule HCl, for example, the hydrogen atom has a charge of +0.18 and the chlorine atom a charge of −0.18. The fractional charges are known as **partial charges**. A convenient way of setting up a set of transferable partial charges is to give each atom a contribution to the partial charge from each type of bond that it is involved in. For example, in chloroethane, CH_3CH_2Cl, we need to consider contributions for the carbon atoms for carbon bound to carbon, carbon bound to hydrogen and carbon bound to chlorine. Carbon bound to carbon is given a value of zero. For elements such as oxygen, which can be singly or doubly bonded ($C-O$ or $C=O$), we need different partial charge contributions for each type of bond.

In one available computer program carbon bonded to hydrogen gives a contribution of +0.053.

○ What is the partial charge on carbon in methane using this value?

○ The carbon in methane is attached to four hydrogens so its partial charge is 4×0.053, or 0.212.

○ What is the partial charge on the hydrogen atoms?

○ Since methane is a neutral molecule, the hydrogens must each have a partial charge of −0.053.

QUESTION 2.3

Calculate the partial charges on all carbons in 2,2,4-trimethylpentane, **2.1**, using the value +0.053 for carbon bound to hydrogen.

$$
\begin{array}{ccc}
& CH_3 & CH_3 \\
& | & | \\
H_3C-C & -CH_2-CH-CH_3 \\
& | & \\
& CH_3 & \textbf{2.1}
\end{array}
$$

As well as the electrostatic interaction arising from the partial charges, we also need intermolecular forces. These can be important for large atoms such as bromine or iodine.

These two interactions alone, however, are insufficient, the main problem being that these interactions are the same in all directions, whereas chemical bonding acts predominantly in the direction of the two atoms bonded. We therefore need to add in terms to represent bonding. The various forces used are illustrated pictorially in Figure 2.5.

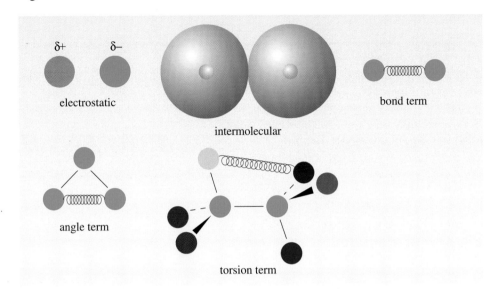

Figure 2.5
Schematic representation of forces used in molecular mechanics of organic molecules.

First we assume that a force similar to that produced by a spring joins the bonded atoms, the bond term. It requires energy to pull the atoms further apart and to push them closer together than the equilibrium bond distance, just as you need to supply energy to stretch or compress a spring.

Then a spring-like force is also introduced to preserve bond angles, the angle term. The energy increases as the angle increases or decreases from its equilibrium value. This is similar to the term we had to add for zeolites to preserve $\angle OSiO$ and $\angle OAlO$ angles.

Finally, we have to take into account the change in energy due to the twisting of one bond relative to another, for example the rotation of a CH bond in one CH_3 group of ethane relative to a CH bond in the other CH_3 group. The relative orientations of the two bonds are given by the **torsion angle**. This torsion term will tend to conserve the lowest energy conformation. A torsion angle between two bonds C—X and C—Y joined by a C—C bond is defined as the angle between the XCC plane and the CCY plane. It is perhaps easier to understand if we use a Newman projection. In a Newman projection you are looking along the C—C bond, and the angle between the two planes is given by the angle between C—X and C—Y in this projection (if X and Y are eclipsed, this angle should be taken as $0°$, although it is customary to show C—X and C—Y slightly separated for clarity).

Structure **2.2** shows ethane in the eclipsed conformation as a Newman projection:

The torsion angle H_1CCH_5 is $60°$, that is, the angle between the H_1C and CH_5 bonds in this projection.

🔵 What is the H_1CCH_6 torsion angle?

⚪ $180°$.

H_1
H_4 H_5

H_3 H_2
H_6

2.2

● The bond terms needed for ethane are C—H and C—C. Are any extra bond terms needed for butane, C_4H_{10}?

● No. Bond terms are transferable, and butane only contains C—C and C—H bonds.

● Are any extra angle terms needed for butane?

● Yes a ∠CCC angle term.

Several computer programs are available that include one or more sets of parameters for all functional groups commonly found in organic chemistry. These are usually obtained using a training set of molecules. A **training set** consists of a group of small molecules that incorporate most of the common functional groups and the more usual ring systems. The parameters are varied until a good fit between experimental and calculated geometries is obtained. This ensures that for most molecules molecular mechanics gives a good approximation to the experimental geometry.

As there are different methods of estimating all the terms, such packages may give you a choice of data set to use.

One common use of such packages is to find minimum energy structures of molecules. To obtain the minimum energy structure of a molecule you first have to draw a rough structure on the computer screen. Many packages provide fragments such as —COOH groups that you can simply join together to give a first guess. You then have to choose a data set.

Having decided on a set of data, you can then ask the program to find the lowest-energy geometry. It will do this by moving the atoms around, calculating the energy at the new positions and then moving them again until it reaches a geometry from which movement in any direction leads to an increase in energy. This process of finding the lowest-energy geometry is known as **minimization**. Several ingenious methods of deciding how to move the atoms have been devised.

On the CD-ROM you will see the results of such minimization for different molecules. Figure 2.6 shows aspirin, the common painkiller, in the geometries obtained by minimization using two molecular mechanics data sets compared with the experimental geometry. The sets are called M and S. Data set S (Figure 2.6b) is a very basic set whereas data set M (Figure 2.6c) was obtained using a large set of molecules which gives more accurate geometries and conformations for organic molecules and biopolymers. Note, for example, that the orientation of the CH_3COO- group using set S differs from that in the experimental geometry and as predicted by set M.

COMPUTER ACTIVITY 2.1

Now would be a good time to study the molecular mechanics section of *Molecular Modelling* on the CD-ROM associated with this Book.

This sequence illustrates the results of some calculations on molecules and solids using molecular mechanics.

QUESTION 2.4

List all the bond terms you would need to describe the cocaine analogue ecgonine, **2.3**.

2.3

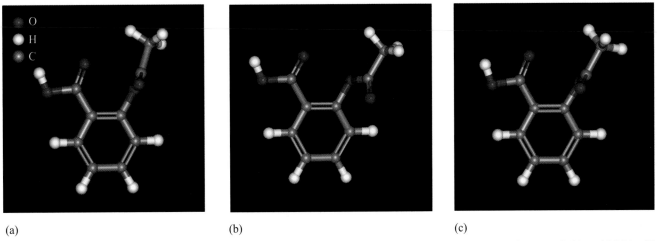

(a)   (b)   (c)

Figure 2.6 Aspirin in its experimental geometry (a) and as minimized by molecular mechanics, using data sets S (b) and M (c). 🖥

QUESTION 2.5

The optimized structure of butadiene ($H_2C=CH-CH=CH_2$) in a *cis-type* conformation obtained from the two molecular mechanics data sets M and S is given in Structure **2.4**.

2.4

What might cause the bond lengths for the two sets to differ?

2.3 Summary of Section 2

1   Molecular mechanics is based on a simplified model of bonding but can produce useful results for large molecules, such as proteins and solids.

2   For ionic solids, the model is of ions held together mainly by electrostatic interaction with some contribution from intermolecular forces.

3   Aluminosilicates such as zeolites are more sensitive to intermolecular forces between ions and in addition require a term that retains the ∠OSiO and ∠OAlO angles.

4   Modelling of organic molecules uses partial charges rather than the formal charges of the ionic model.

5   The most important terms for organic molecules are those that use spring-like functions to retain the bond lengths, bond angles and torsion angles.

QUANTUM CHEMISTRY OF ATOMS

3

Molecular mechanics has proved very useful especially for large molecules, where more exact methods would require impractical amounts of computational resource. There are problems that molecular mechanics cannot tackle, however, for example the nature of transition states in chemical reactions or magnetic and spectroscopic properties of molecules. Even for such a simple molecule as dioxygen, O_2, we need quantum theory to explain why it has two unpaired electrons.

Quantum theory was first put forward at the beginning of the twentieth century and the equation describing its application to atoms and molecules was proposed by Erwin Schrödinger (Box 3.1) in 1926. It was not until the 1950s, however, that it began to be used widely by chemists. This was because, although it was possible to write down the Schrödinger equation for any system, solving the equation for molecules of chemical interest was impractical before the advent of computers. The advances in computer software and hardware since then have been such that by 1998, experimental chemists routinely used quantum calculations performed on a desktop computer as a complement to their work. The Nobel Prize for Chemistry in that year was awarded to two pioneers of quantum chemistry — John Pople and Walter Kohn (Box 3.2).

BOX 3.1 Erwin Schrödinger

Erwin Schrödinger was born in 1887 near Vienna into a family where both English and German were spoken. At school he showed a gift for understanding maths and physics as soon as it was taught without doing any homework. After obtaining his doctorate in 1910 he took up several academic posts but it was in Zurich, where he was appointed professor of theoretical physics in 1921, that he found the right intellectual atmosphere. It was here that he produced his equation describing the quantum theory of atoms and molecules. In 1933 he moved to Oxford. He was offered a chair at Edinburgh but due to Home Office delays went to Graz in Austria. In 1938 he was dismissed for 'political unreliability'. With the help of de Valera, then President of the Assembly of the League of Nations, he obtained a post at Dublin where he remained until he retired.

Figure 3.1
Erwin Schrödinger.

BOX 3.2 John Pople and Walter Kohn

John Pople was born in Burnham-on-Sea, Somerset in 1925, the son of a men's clothing storeowner. His parents considered education important and sent him to Bristol Grammar School. Here he developed an interest in mathematics and in 1943 went to Trinity College, Cambridge to read mathematics. As it was wartime he had to finish his degree in two years and in 1945 went to work for the Bristol Aeroplane Company. In 1947 he returned to Cambridge as a maths student but developed an interest in theoretical science. He went on to work with Sir John Lennard-Jones after whom one of the interatomic potentials used in calculations is named. At about the same time Pople decided to learn to play the piano and went on to marry his piano teacher, Joy Bowers.

During the ten years that Pople spent in Cambridge he was in contact with many famous researchers who were active in Cambridge at the time. This was the time when the first crystal structures of proteins were being solved, the double helix nature of DNA was proposed and major cosmological theories were put forward. Pople's work in chemistry began with a study of the non-bonding pairs of electrons in water. In 1955 he developed an interest in NMR spectroscopy and moved to Ottawa. He then moved to the National Physical Laboratory in Teddington, London but found that the administrative burden hampered his research.

It was after moving to Pittsburgh, USA in 1964 that John Pople began the work in computational chemistry that won him the Nobel Prize. He remained there until 1993.

Walter Kohn was born in Austria in 1923 to middle-class Jewish parents. His father owned a small business. His mother was highly educated with a knowledge of a wide range of languages. He attended the Akademische Gymnasium in Vienna where his interests lay more in Latin than in maths and science. In 1938, however, following the annexation of Austria by Hitler, his father's business was confiscated and he was expelled from school. He moved to a Jewish school where he was inspired by teachers in physics and maths and then emigrated to England without his parents who could not leave Austria. After school he moved to the University of Toronto. His parents and maths and physics teachers perished in the Holocaust.

At Toronto he studied maths and physics, not being allowed into the chemistry department due to his nationality, as war work was in progress there. Among his fellow students was Arthur Schawlow who later shared a Nobel Prize for the development of lasers.

He moved to Harvard to take his Ph.D. and from there moved to the Carnegie Institute of Technology in Pittsburgh to teach solid-state physics. From 1953 Kohn had a series of summer jobs at Bell Laboratories (at that time the outstanding centre for solid state research) where he worked with Shockley, the inventor of the transistor.

In 1960 Kohn moved to California and it was here that he developed the density-functional theory for which he was awarded the Nobel Prize.

(a)

(b)

Figure 3.2
(a) John Pople and
(b) Walter Kohn.

Quantum theory is based on the concept of an electron behaving as a wave, and one of the consequences of this is that we cannot picture the electron as moving around the nuclei in a definite path like the orbit of a planet around the Sun. We can, however, say how likely the electron is to be at any particular point in space. When we observe a molecule experimentally, for example by X-ray diffraction, there is a higher density of electrons at some points than at others. A way of picturing this is to imagine the electrons smeared out over the atom or molecule, more thickly spread in some places than others. In Figure 3.3, for example, we have pictured an electron in a hydrogen atom; the areas where the dots are closer together are those where the electron is more likely to be found.

If we now add another hydrogen nucleus to form H_2^+, the electron distribution changes. The electron now is likely to be near either of the two nuclei (Figure 3.4). In addition, the electron is more likely to be found between the nuclei than it would be if the distribution around each nucleus was the same as in Figure 3.3.

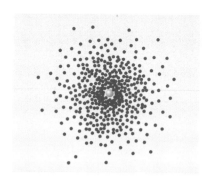

Figure 3.3
Superimposed imaginary photographs of an electron in a hydrogen atom.

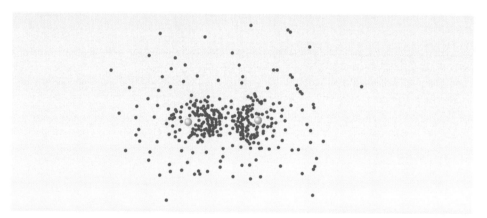

Figure 3.4
Superimposed imaginary photographs of an electron in H_2^+.

Because Schrödinger's equation describes an electron as wave-like, the solutions to this equation are called **wavefunctions**. Consider for a moment a stretched piece of string or a violin string that is plucked. At different points along the string the string is displaced from its original position by a different amount. A snapshot at one instant in time might make the string appear as in Figure 3.5.

Under certain conditions we can set up a standing wave in the string. Although the displacement of the string varies with time, in a standing wave the maximum displacement is always at the same point or points along the string. Figure 3.6 shows such a standing wave at two different times.

The variation of the displacement with distance along the string can be described by a mathematical expression. This is the wavefunction for this standing wave.

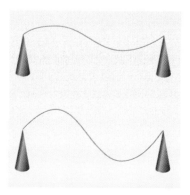

Figure 3.6
A standing wave in a string at two different times.

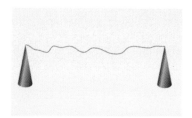

Figure 3.5
A plucked string showing displacements along the string at an arbitrary time.

The wavefunction for an electron in an atom or molecule is a function for a standing wave that describes how the electron is distributed in space. For the wave in the string we only had to consider one dimension, the distance along the string, but for the electron wave we have to deal with three dimensions. From the wavefunction of the electron we can calculate properties of atoms and molecules. One particularly important quantity is the **electron density**. According to Schrödinger's theory, the electron density at any point in space for an atom or molecule is given by the square of the electron wavefunction for the atom or molecule at that point. The electron density can be interpreted as the probability of finding an electron at any point and this leads to the electron wave being referred to as a probability wave.

Quantum theory enables us to calculate a wavefunction for electrons in atoms and molecules and, from this, the energy and other properties of molecules. Apart from some small and not particularly interesting molecules such as H_2^+, however, we cannot obtain a mathematical function as an exact solution to the Schrödinger equation although modern computational methods can give a solution as a series of numbers that give an electron distribution very close to the experimental one. For general calculations on molecules, therefore, chemists have developed approximate solutions, some of which give answers that agree remarkably well with experiments.

One common approximation made is that the wavefunction for an electron in a molecule can be built up by combining the wavefunctions for electrons in the constituent atoms. This has an attraction for chemists as we can relate the results to our pictures of atoms connected by chemical bonds. We start this Section then by looking at atomic wavefunctions.

3.1 Atomic orbitals

When the Schrödinger equation is solved for the hydrogen atom, it is found that only certain standing wave solutions exist. These solutions for one electron are known as **orbitals**, a term that derives from an earlier model of the atom in which electrons occupied orbits around the nucleus as the planets do around the Sun. If we work out the energies of these orbitals for the hydrogen atom, we find they correspond to the experimentally found energy levels $n = 1$, $n = 2$, $n = 3$, etc. The distribution in Figure 3.3 is that of an electron in the 1s orbital (that is, one with $n = 1$ and $l = 0$). This electron has a spherical distribution in space; it is equally likely to be found in any direction. However, its distribution varies with distance from the nucleus. The electron is much more likely to be close to the nucleus than far away.

One way of representing the electron distribution is as shown in Figure 3.7 (overleaf). The likelihood of an electron being at a point in the plane is represented by the height of the plot above the point.

Another way of representing such distributions is by means of a contour diagram. In this, we draw a line joining all points where the electron is equally likely to be. Such a diagram is shown for the 1s orbital in Figure 3.8 (overleaf). The contours close to the nucleus join points of high probability of finding the electron; those further away correspond to a lower probability. If you think of Figure 3.7 as a hill, then Figure 3.8 would be the contour map of that hill.

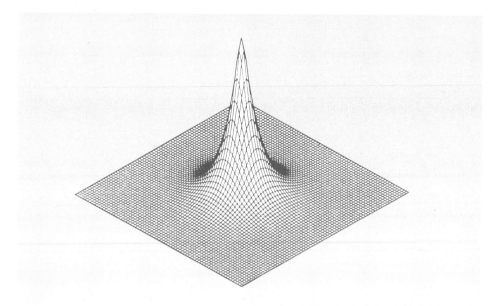

Figure 3.8
Contour diagram for a 1s electron in a hydrogen atom. The numbers denote electron density.

Figure 3.7 Computer plot of a 1s electron in a hydrogen atom. The probability of being at any point in a plane is represented by a height above the plane. The peak is above the nucleus.

A quicker way of representing the electron is to pick just one contour. The one chosen is usually one such that the electron has a 95% chance of being inside it. For a 1s electron in hydrogen this repesentation is shown in Figure 3.9. Figures 3.7 to 3.9 represent the electron probability in a plane but, of course, the electron is free to move in three dimensions. Figure 3.10 shows the 95% contour in a 3-D representation. This contour is referred to as a **boundary surface**.

In general, if you need to draw an orbital, we suggest you use the single contour as in Figure 3.9 because this is more easily remembered and drawn freehand.

So far we have only considered the 1s electron. How does the distribution vary if the electron is not in the 1s level but in a 2s or 2p or higher energy level? The boundary surfaces for electrons in 1s, 2s, 2p, 3s, 3p and 3d orbitals in hydrogen are shown in

Figure 3.9
The 95% probability contour for a 1s electron in a hydrogen atom.

Figure 3.10
The 95% probability contour for a 1s electron in a hydrogen atom in 3-D, a boundary surface.

Figure 3.11 (overleaf). Note that the distribution of the s electrons (that is, those with $l = 0$) is spherical, that of the p electrons ($l = 1$) forms two separate lobes, and that of the d electrons ($l = 2$) forms four lobes (with the exception of $3d_{z^2}$, which forms three). The p and d distributions are not spherical; an electron in one of these is more likely to be found along some directions than others. Where the contour is in several parts, the line or surface surrounding each part represents the same value of the electron distribution. The situation is rather like a mountain with twin peaks. If we picked a contour corresponding to a height above the point where the peaks split, then on a map we would see two separate lines, one surrounding each peak.

⬤ There is only one 1s orbital. How many 2p orbitals would you expect there to be from your knowledge of electron configurations?

⬤ Three. There was one box for the s sub-shell but three for the p sub-shell.

Each of the three p orbitals has its lobes along the x, y or z direction. Thus we can label them p_x, p_y or p_z. An electron occupying a $2p_x$ orbital, for example, is more likely to be found along the x axis than along the y or z axes. Note, however, that if an atom has one p electron, then that electron does not remain exclusively in $2p_x$, $2p_y$ or $2p_z$, but spends an equal amount of time in all three. Thus if we excited the electron in a hydrogen atom to the 2p level and then measured the electron distribution we should find that it was spherical — the result of combining the $2p_x$, $2p_y$ and $2p_z$ distributions.

If we take a section through the boundary surfaces in Figure 3.11, then we obtain the 95% contour diagrams shown in Figure 3.12 (p. 27). Note that we have only shown representative boundary surfaces for 2p, 3p and 3d. The remaining surfaces will be similar to the $2p_z$, $3p_z$ and $3d_{xz}$ shown, but with different orientations.

The figures we have drawn so far show only how much electron there is at any point, but when we come to combine atoms to form molecules, another property of the electron wave is also important. This is the phase or sign. Figure 3.13 (p. 28) shows some simple waves. In Figure 3.13a the two waves go up and down at the same time: they are said to be **in phase**. If we add these two waves together then we get a bigger wave, which is still moving up and down in the same pattern. In Figure 3.13b one wave goes up as the other goes down: the waves are **out of phase**. If these two waves are added together, they cancel each other out.

The orbitals describing electrons in atoms or molecules are more complex than the simple waves in Figure 3.13, but when they combine in phase or out of phase the results are similar.

The boundary surfaces in Figure 3.11 are for the electron density, the probability of an electron being at any point in space. The electron density is given by the square of the wavefunction. Points with the same electron density will have the same numerical value for the wavefunction, but the wavefunction may be positive or negative. For example, if the probability of finding an electron at a particular point was one-quarter, 0.25, then the wavefunction at that point would have the value plus one half, $+ 0.5$, or minus one half, -0.5, since both $(0.5)^2$ and $(-0.5)^2$ are equal to 0.25. The sign of the wavefunction gives its phase. To represent the wavefunction itself we can use the same contours as for electron density, but we also need to indicate the phase of the wavefunction. In atomic and molecular orbital representations, we shall use colour to show differences in phase. 1s orbitals are all one phase and so are shown in one colour. 2p orbitals have two lobes, which are out

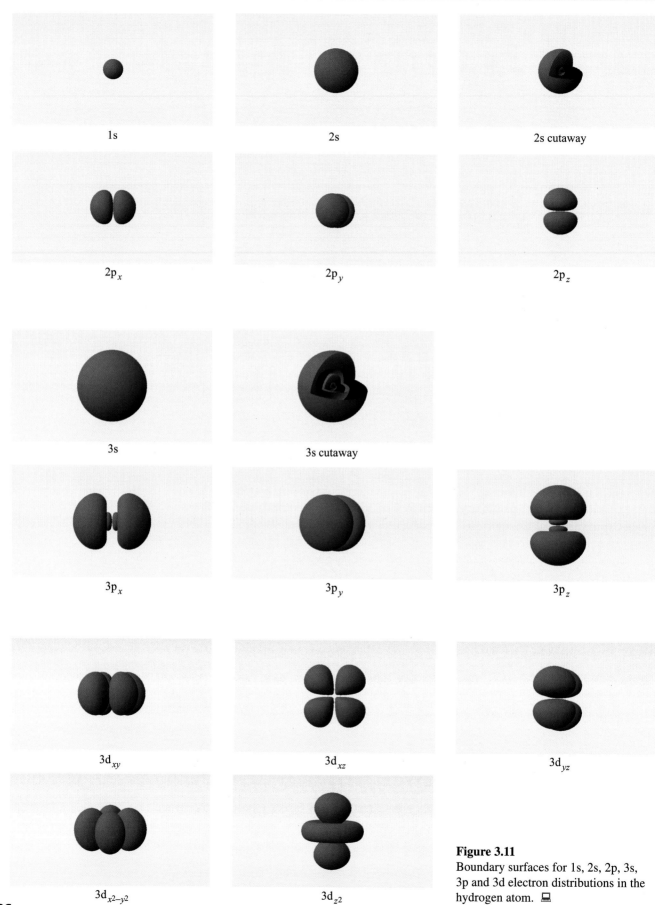

1s

2s

2s cutaway

$2p_x$

$2p_y$

$2p_z$

3s

3s cutaway

$3p_x$

$3p_y$

$3p_z$

$3d_{xy}$

$3d_{xz}$

$3d_{yz}$

$3d_{x2-y2}$

$3d_{z2}$

Figure 3.11
Boundary surfaces for 1s, 2s, 2p, 3s, 3p and 3d electron distributions in the hydrogen atom.

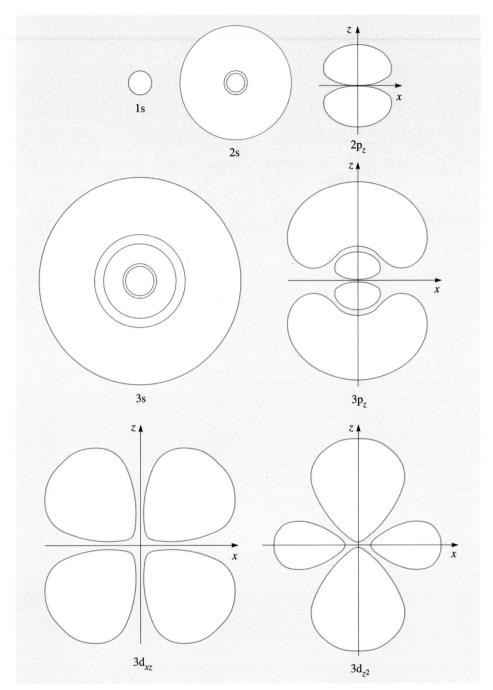

Figure 3.12 The 95% contours for 1s, 2s, 2p, 3s, 3p and 3d electrons in the hydrogen atom. *Note*: only representative orbitals are shown for 2p, 3p and 3d.

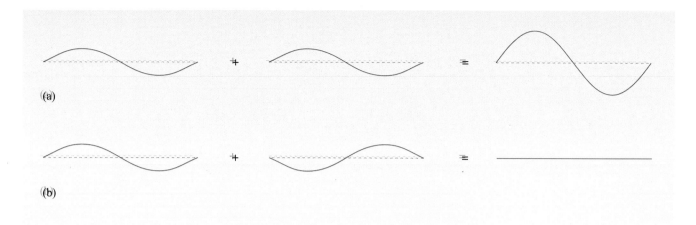

Figure 3.13 Addition of simple waves in phase (a) and out of phase (b).

of phase with each other, and we illustrate this by showing them in different colours. The four lobes of the 3d orbital alternate in phase as you move around the atom.

Figure 3.14 shows the same orbitals as in Figure 3.12 but coloured to indicate parts of the orbital that differ in phase. The different phases can also be denoted by + and − signs, and these are also shown in Figure 3.14. The + and − refer to the value of the wavefunction; in the regions labelled + the mathematical function describing the electron wave is a positive (fractional) number, whereas in the regions labelled − it is a negative number. (*Note*: + *and* − *do not represent electrical charge*.) Boundary surfaces of hydrogen orbitals are shown in Figure 3.15 (p. 30), where we have included cutaway diagrams for 2s and 3s so that you can see all the surfaces corresponding to the contours in Figure 3.14. It is impossible to tell the phase of an orbital in an isolated atom but, as you will see in Section 4, when atoms combine to form molecules, the relative phase of orbitals on the different atoms is very important.

3.1.1 Orbitals of atoms other than hydrogen

The starting point for the calculation of wavefunctions for other atoms is generally to assume that each electron is in an orbital like the hydrogen orbitals. Such orbitals are known as **atomic orbitals**.

You will probably be aware in describing electronic configurations of atoms, that two electrons could occupy the 1s level.

● What property differs for these two electrons?

● They have to have opposite spin.

● How many electrons can occupy the 2p level?

● Six. Each of the three 2p orbitals can contain two electrons.

In the same way, when building up the atomic wavefunction, we assign up to two electrons of opposite spin to each orbital.

We assign electrons to atomic orbitals starting from the lowest energy orbital. Two electrons, of opposite spin, can go into each orbital and, where there are two or more

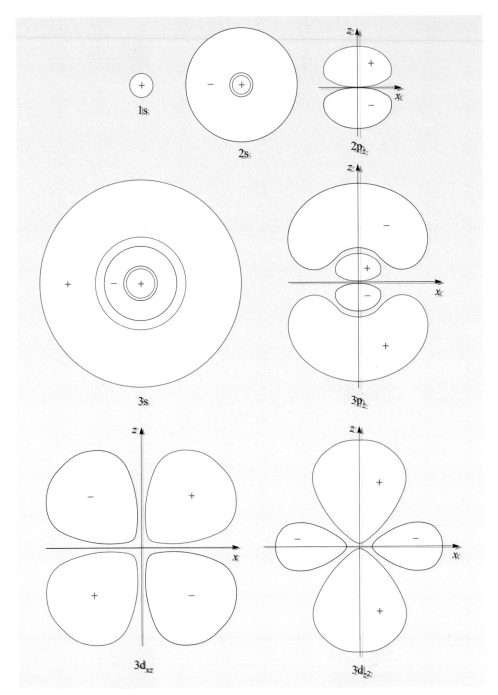

Figure 3.14 The 95% contour for 1s, 2s, 2p, 3s, 3p and 3d orbitals of the hydrogen atom, with the phases denoted by + and − signs. *Note*: only representative orbitals are shown, as in Figure 3.12.

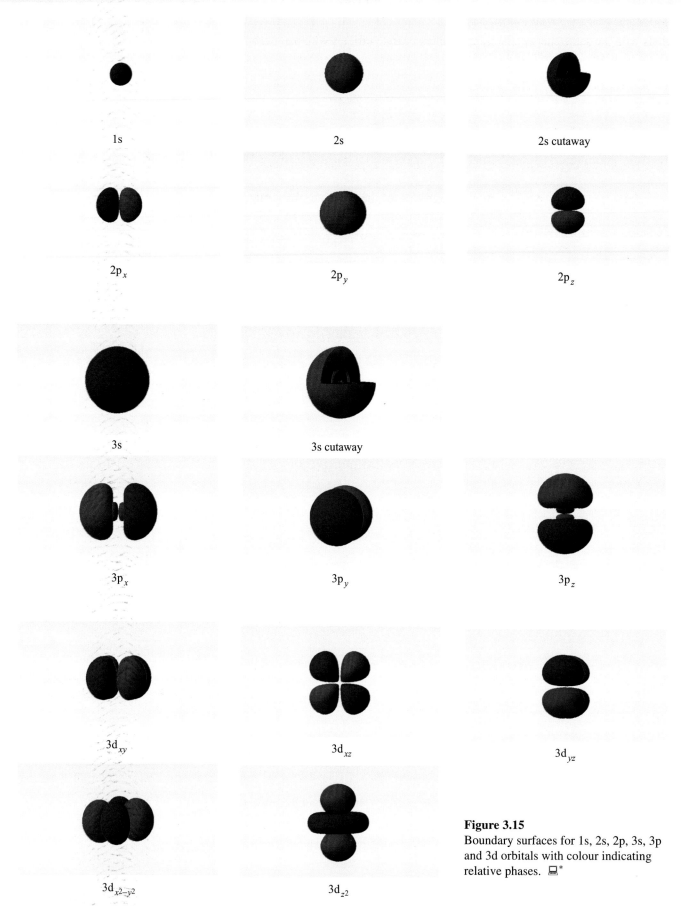

1s

2s

2s cutaway

$2p_x$

$2p_y$

$2p_z$

3s

3s cutaway

$3p_x$

$3p_y$

$3p_z$

$3d_{xy}$

$3d_{xz}$

$3d_{yz}$

$3d_{x2-y2}$

$3d_{z2}$

Figure 3.15
Boundary surfaces for 1s, 2s, 2p, 3s, 3p and 3d orbitals with colour indicating relative phases. 🖥*

orbitals with the same energy, electrons go into separate orbitals with parallel spins first (Hund's rule).

So, going across the second row of the Periodic Table, in lithium we put two electrons into 1s and one into 2s ($1s^22s^1$), in beryllium we put two electrons into each of 1s and 2s ($1s^22s^2$) and at boron we add an electron to a 2p orbital ($1s^22s^22p^1$) (Figure 3.16). At carbon we have to put two electrons into 2p orbitals. There are three 2p orbitals all of the same energy, but electrons with parallel spins tend to keep apart, and so putting the 2p electrons in with parallel spins reduces electron repulsion. Rather than put two electrons of opposite spin into one 2p orbital, therefore, we place the electrons one in each of the two 2p orbitals with parallel spins ($1s^22s^22p^2$). For nitrogen we have two paired electrons in 1s, two paired electrons in 2s and three electrons with parallel spins in the three 2p orbitals ($1s^22s^22p^3$). At oxygen we have to start pairing 2p electrons so we have a pair in 1s, a pair in 2s, a pair in one of the 2p orbitals and two single electrons with parallel spins in each of the other two 2p orbitals ($1s^22s^22p^4$). Fluorine has two paired 1s electrons, two paired 2s, two 2p orbitals each with a pair of electrons and one 2p orbital with one electron ($1s^22s^22p^5$). Finally, at neon, the 1s, 2s, and all three 2p orbitals contain pairs of electrons ($1s^22s^22p^6$).

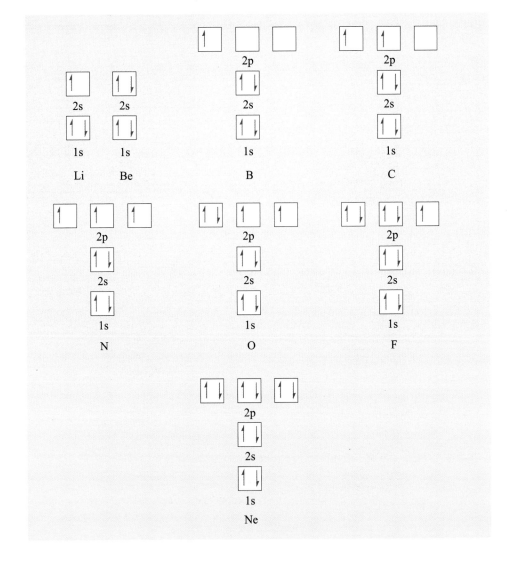

Figure 3.16
Electronic configurations for the atoms Li to Ne.

These orbitals are not the same as the corresponding orbitals in hydrogen. Because all atoms are spherical, the orbitals can still be labelled s, p, d, etc. and so have the same shape, but the average distances of the electrons from the nucleus are different from those in hydrogen. Figure 3.17 shows the 1s orbital in lithium compared with that in hydrogen. The orbital in lithium is concentrated closer to the nucleus because of the increased electrostatic interaction due to the increased nuclear charge, which is not entirely offset by the repulsion from the other electrons. This difference is illustrated in another way in Figure 3.18, where we show the way the 1s orbital varies with distance from the nucleus for H and Li. Here you can see that the H 1s orbital is more spread out. For each atom, electrons in orbitals of higher principal quantum number n are, on average, further out from the nucleus than those of lower value of n. So the 2s electron in lithium, although closer to the nucleus on average than a 2s electron in an excited H atom, will be farther from the nucleus than the 1s electron in lithium. Similarly, an electron in a 3s orbital in sodium will, on average, be further from the nucleus than electrons in 2s and 2p orbitals in sodium.

The size of an atom or ion is determined by the electron density distribution of its outermost orbital(s). As we go down a Group of the Periodic Table — that is, as n increases — the radii of the atoms and their ions increases.

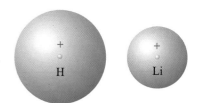

Figure 3.17
1s orbitals in H and Li.

- What does this tell us about the electron density in the outermost orbitals as you go down a Group of the Periodic Table?

- The electron density becomes significant further out from the nucleus; the electron is on average further from the nucleus.

So an electron in a 3p orbital in argon will be closer to the nucleus than one in a 3p orbital in an excited neon atom, but further from the nucleus than an electron in a 2p orbital in neon.

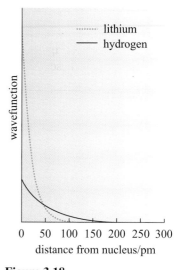

Figure 3.18
Variation of 1s orbitals in H and Li with distance from the nucleus.

QUESTION 3.1

How will a 1s orbital in sodium compare with that in lithium? How will a 3s orbital in sodium compare with a 2s orbital in lithium?

3.1.2 Summary of Section 3.1

1 Standing wave solutions to the Schrödinger equation for an electron in an atom or molecule are known as orbitals.

2 The Schrödinger equation for the hydrogen atom can be solved exactly and the orbitals obtained can be linked to the levels 1s, 2s, 2p... found experimentally for the hydrogen atom.

3 Orbitals of different second quantum number, l, have different directional properties, e.g. all s orbitals are spherical.

4 Orbitals of atoms other than hydrogen have the same directional properties corresponding to s, p, d..., but the variation of the orbital with distance from the nucleus differs.

5 On average, electrons with higher principal quantum number, n, are further from the nucleus.

6 On average, an electron in a particular orbital, e.g. 2s, will generally be closer to the nucleus the greater the nuclear charge of the atom.

7 When building up a wavefunction for an atom with more than one electron, the electrons are assigned to atomic orbitals starting from that of lowest energy.

8   Two electrons of opposite spin can be assigned to each orbital. For orbitals of
the same energy, electrons are put first into separate orbitals with parallel spin
according to Hund's rule.

3.2 Calculating atomic orbitals

The way an atomic orbital varies with direction is known exactly, giving the shape
of s, p, d, etc., orbitals. For atoms with many electrons, however, the way the orbital
varies with distance from the nucleus can only be approximated, albeit very
accurately with modern computational methods. Most calculations start with a set
of mathematical functions representing the orbital. This set is called a **basis set**.
Basis sets have been developed by many people, the most famous being John Pople.
In nearly all modern calculations these basis sets are sums of terms known as
Gaussian functions. Figure 3.19 shows several Gaussian functions for the 1s orbital
of fluorine. One of these functions alone is not a very good approximation to an

Figure 3.19
Six Gaussian functions used to
describe the 1s orbital in fluorine.

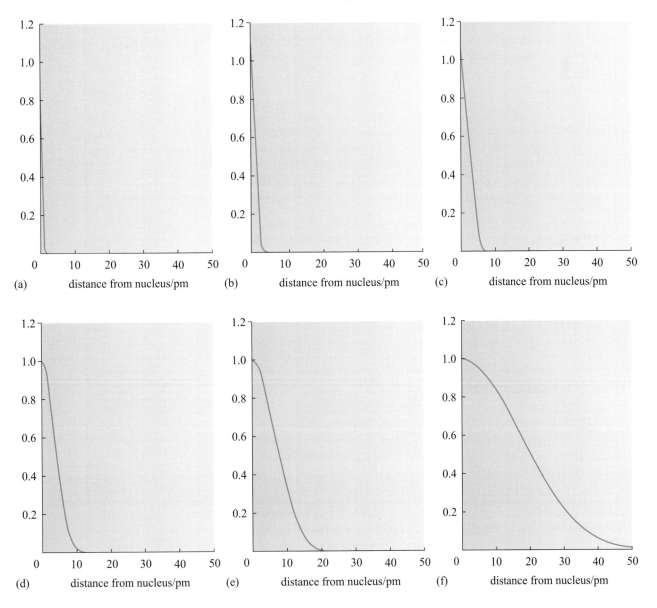

atomic orbital, but by adding different amounts of several together we can improve the fit. Figure 3.20 is a representation of a 1s orbital for fluorine formed by adding together different amounts of the six Gaussians shown in Figure 3.19. The individual Gaussians are flatter near the nucleus than an actual orbital; the addition as in Figure 3.20 sharpens the curve near the nucleus.

A possible basis set for fluorine might include six Gaussians for the 1s orbital, four for the 2s orbital and four for the 2p orbitals. Note that only one set of Gaussians is needed for all three 2p orbitals, as the Gaussians describe the variation of the wavefunction with *distance* from the nucleus, and $2p_x$, $2p_y$ and $2p_z$ vary only *in direction*. In general an atom will have a set of Gaussians for each occupied orbital but for more accurate calculations additional sets may be added.

- Figure 3.20 is for a 1s orbital. How would you expect the proportions of different Gaussians to vary for a 2s orbital?

- A 2s electron is on average further from the nucleus than a 1s electron and so would need more of the wider Gaussians as in Figure 3.19f, and fewer of the narrower ones as in Figure 3.19a and b.

Figure 3.21 shows three Gaussians used to represent the 4s orbital in potassium (a–c), and the sum of these Gaussians (d). Note how much further from the nucleus the Gaussians in (d) spread.

Gaussians were originally introduced because they were easier to deal with mathematically than other functions that fitted the orbitals more accurately. They continue to be used because a good approximation to orbitals can be obtained from a reasonably small set. Orbitals of valence electrons (valence orbitals) are often well described by three to five Gaussians. Different sized basis sets containing different numbers of Gaussians are available and can be chosen to suit the problem you want to solve.

In the standard procedure, known as the **Hartree–Fock** (HF) method, these atomic basis sets are used to form the initial guess orbitals, and electrons are fed in using Hund's rule. For each electron in turn, its energy in an orbital is calculated assuming it is moving in the average electron distribution produced by other electrons occupying their orbitals. The orbital is then varied and the energy recalculated until a minimum energy is reached. This process is then repeated until no further change is produced.

3.3 Hybrid orbitals

The first quantum chemical theory to give a convincing explanation of the shapes of molecules was the concept of **hybrid orbitals**. This is best illustrated by using carbon as an example, and is appropriate since the language of hybridization is still commonly used in organic chemistry.

Carbon has the electronic configuration $1s^2 2s^2 2p^2$ with two unpaired electrons, and yet we know that carbon in its compounds almost invariably forms four bonds. The proposed solution to this was to reorganize the 2s and the three 2p orbitals to form four equivalent *hybrid orbitals*. If this is done, four identical orbitals directed towards the corners of a tetrahedron are produced. The formation of just one of these orbitals from a 2s and three 2p orbitals is shown in Figure 3.22. With one

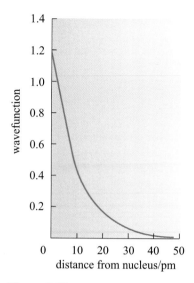

Figure 3.20
The Gaussians in Figure 3.19 added together to form a representation of the 1s orbital.

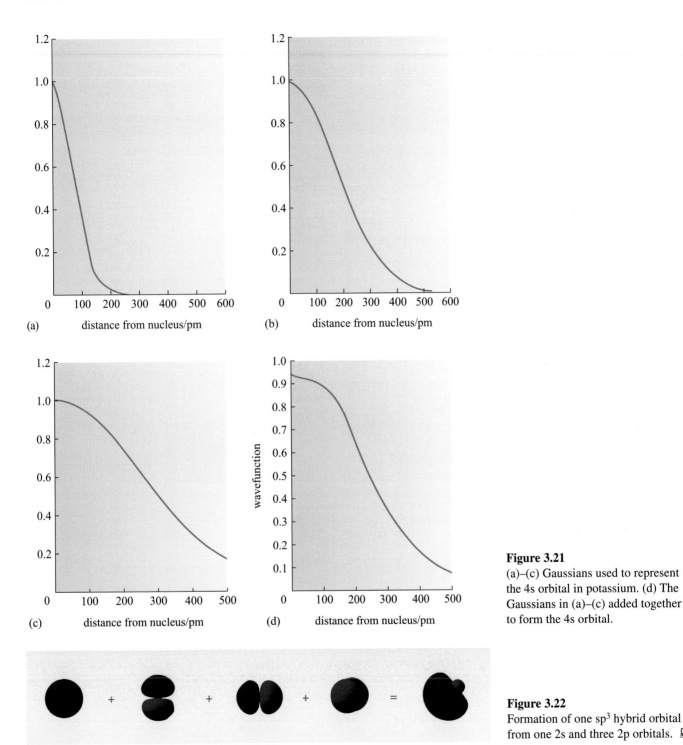

Figure 3.21
(a)–(c) Gaussians used to represent the 4s orbital in potassium. (d) The Gaussians in (a)–(c) added together to form the 4s orbital.

Figure 3.22
Formation of one sp³ hybrid orbital from one 2s and three 2p orbitals. 🖳

electron in each we are set up to form four single bonds as in methane. The hybrid orbital in Figure 3.22 is known as an sp³ hybrid.

Carbon compounds with double bonds (C=X, where X = C, O, N or S) have a planar rather than a tetrahedral geometry. For carbon double bonds we leave one 2p orbital as it is and only hybridize the other two with the 2s orbital. This gives us an sp² hybrid (Figure 3.23 overleaf). In ethene there will be one electron in each sp² hybrid and one in the non-hybridized p orbital on each carbon. Two of the electrons

Figure 3.23
Formation of one sp² hybrid orbital
from one 2s and two 2p orbitals.

in sp² hybrids will form bonds to hydrogen and one will form a bond to the other carbon. The electron in the p orbital will also form a bond to the other carbon, giving a carbon–carbon double bond. We shall return to the p orbital not involved in hybridization later.

◉ What will be the hybridization of the two carbons in ethanal, Structure **3.1**?

◉ The carbon of the CH₃ groups will be sp³ hybridized and that of the C=O group sp² hybridized.

(Structure 3.1)

$$H_3C-C\overset{\displaystyle O}{\underset{\displaystyle H}{\diagup\diagdown}}$$

3.1

In ethyne with its triple bond (HC≡CH), two p orbitals on each carbon are left as p orbitals and two linear sp hybrids are formed (Figure 3.24). The electrons in the sp hybrids form the carbon–hydrogen bonds and a carbon–carbon bond. The remaining two electrons on each carbon form carbon–carbon bonds as in ethene.

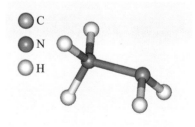

Figure 3.24
Formation of one sp hybrid orbital
from one 2s and one 2p orbital.

Thus if carbon forms four single bonds, it is sp³ hybridized; if it forms a double bond it is sp² hybridized and if it forms a triple bond it is sp hybridized.

Although hybridization is most commonly used in organic chemistry, it is not confined to carbon. Consider ammonia, NH₃. In ammonia, nitrogen forms three single bonds to hydrogen, leaving a non-bonding pair of electrons in the valence shell. Where we have non-bonding pairs, we assume they occupy a hybrid orbital as this assumption leads to prediction of the correct geometry. So for nitrogen in ammonia we need four hybrid orbitals; that is, the nitrogen is sp³ hybridized. One hybrid orbital can be filled with two electrons, leaving the other three to form bonds. Thus ammonia will have three N—H bonds arranged tetrahedrally and a fourth tetrahedral position occupied by a non-bonding pair. In methylamine, CH₃NH₂, the carbon will be sp³ hybridized with three hybrid orbitals forming C—H bonds and one forming the C—N bond. The nitrogen will also be sp³ hybridized with two hybrids forming N—H bonds, one the C—N bond and the fourth containing a non-bonding pair of electrons. This enables us to predict correctly that the structure of methylamine will have a pyramidal arrangement around nitrogen (Figure 3.25).

◉ Does this theory explain the shape of the water molecule?

◉ Yes, if we assume oxygen forms four sp³ hybrids which are orientated tetrahedrally (two form O—H bonds and two contain non-bonding pairs).

Figure 3.25
The structure of methylamine.

● What will be the geometry around oxygen in dimethyl ether, CH_3OCH_3?

● As in water, the oxygen forms four sp^3 hybrids. Two of the hybrids form $O-C$ bonds and there are two non-bonding pairs, so the COC linkage is V-shaped (Figure 3.26).

As for carbon, if nitrogen or oxygen are involved in double or triple bonds, they become sp^2 or sp hybridized, respectively.

● What is the hybridization of oxygen in methanal, Structure **3.2**?

● sp^2. The oxygen forms a double bond and so is sp^2 hybridized.

H
\
   C=O
/
H

3.2

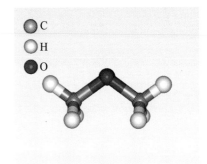

Figure 3.26
The structure of dimethyl ether. ▭

In methanal one electron on oxygen will occupy an sp^2 orbital and forms a bond with an electron in an sp^2 orbital on carbon. There will be two pairs of non-bonding electrons, which will form a trigonal planar arrangement with the CO bond and there will be one electron in a p orbital.

QUESTION 3.2

What is the hybridization of the carbon atoms in (i) HCN, (ii) CH_3COOH, (iii) CH_2OHCH_2OH, (iv) $H_2C=CH_2$, (v) $CH_3CH_2C=NCH_3$.

3.3.1 Summary of Section 3.3

1   The shapes of many molecules, particularly those containing carbon, can be correctly predicted by assuming that ns and np orbitals are reorganized to form hybrid orbitals.

2   Hybrid orbitals formed by combining an s orbital and all three p orbitals are termed sp^3, those involving an s orbital and two of the three p orbitals are termed sp^2, and those formed by combining the s orbital and one p orbital are termed sp orbitals.

3   Carbon, nitrogen and oxygen atoms that are bonded to other atoms purely by single bonds can be regarded as sp^3 hybridized. If one bond is double, the atom is sp^2 hybridized, and if there is a triple bond, the atom is sp hybridized.

MOLECULAR ORBITALS

4

Although we have introduced references to bond formation with hybrid orbitals, we have not yet really tackled how to describe molecules using orbitals. The starting point for molecules, as for atoms, is the Schrödinger equation, and we can solve this to obtain electron wavefunctions or **molecular orbitals**. However, for molecules the electrons are attracted to all the nuclei in the molecule, not just one, and we have to include the repulsion between the nuclei in the energy. There are several methods and many programs available to calculate molecular orbitals. Nearly all employ two approximations.

The first is the **Born–Oppenheimer approximation**. This assumes the nuclei are stationary. Of course, we know that they are not; molecules are continually in motion, rotating and vibrating. However, the light electrons move much faster than the heavier nuclei and so can effectively instantaneously re-adjust as the nuclei move. Assuming that the nuclei are stationary therefore still gives a good approximation for the wavefunction or electron density at the molecular geometry we choose.

The second is to build up the molecular wavefunction (or electron density) from the atomic wavefunctions. In the next Sections, then, we are going to look at how to combine atomic orbitals to give molecular orbitals.

4.1 Orbitals in diatomic molecules

Let us start with the simplest molecule, H_2^+. This has two hydrogen nuclei and one electron. We need to find an orbital that will describe this electron. Since an electron in a hydrogen atom is in a 1s orbital, we make an attempt to approximate the molecular orbital by combining 1s orbitals. The usual way to combine atomic orbitals to make molecular orbitals is to add them as we added the simple waves in Figure 3.13, but in this case we have to add waves on two different centres. So we might try adding together a 1s orbital on one hydrogen nucleus and a 1s orbital on the other hydrogen nucleus. The two 1s orbitals are shown in Figure 4.1. If the two orbitals are in phase as shown in the figure, then where they overlap, the wavefunction and hence the square of the wavefunction increases. This represents an increase of electron density. The overlap lies between the two nuclei and so, in a combination like this, there would be a high electron density between the nuclei. Now this looks useful. If the electron is concentrated between the nuclei, it can act to hold the nuclei together. An electron in such an orbital would be effectively binding the two hydrogen nuclei together to make a molecule.

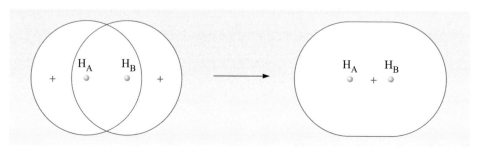

Figure 4.1
Overlap of two in-phase 1s orbitals in H_2^+.

Combining two 1s orbitals in this manner, then, seems a good first approximation to describe bonding using this theory.

We shall label this orbital $(1s_A + 1s_B)$, where $1s_A$ represents a 1s orbital on nucleus A and $1s_B$ a 1s orbital on nucleus B. Its boundary surface is shown in Figure 4.2. If we approximate our atomic orbitals by sets of Gaussians then this molecular orbital is an equal mixture of the Gaussians representing an H 1s orbital centred on nucleus A and the same set centred on nucleus B.

Will the electron in H_2^+ occupy the orbital $(1s_A + 1s_B)$ rather than a 1s orbital on one H atom? Yes it will, because an electron in $(1s_A + 1s_B)$ is more likely to be attracted to both nuclei than one in a 1s orbital. This means that the electron is more strongly bound and its energy is lower. An electron in $(1s_A + 1s_B)$ is therefore of lower energy than one in a 1s orbital. Orbitals such as $(1s_A + 1s_B)$ are called **bonding orbitals**.

Figure 4.2
Boundary surface for the orbital $(1s_A + 1s_B)$. 💻

We can draw a diagram (Figure 4.3) to illustrate the energy advantage of our bonding orbital. In the centre we put boxes or lines to represent molecular orbitals. On either side we put boxes to represent the orbitals on the atoms that make up the molecule. Energy is represented as increasing up the page. Dashed lines join the molecular orbitals to the atomic orbitals that are combined to make them. So, in the centre of Figure 4.3 we have H_2^+ and the orbital $(1s_A + 1s_B)$. On either side we have 1s orbitals. The 1s orbitals are shown higher up the page than $(1s_A + 1s_B)$ because they are higher in energy, and lines join them to $(1s_A + 1s_B)$ because the molecular orbital is a combination of 1s orbitals.

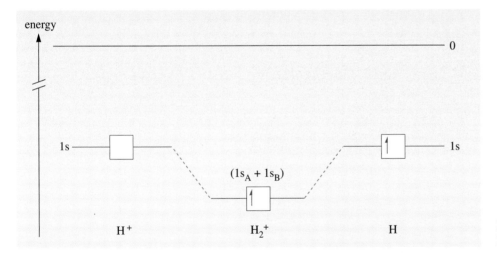

Figure 4.3
Energy-level diagram for H_2^+.

Figure 4.3 shows that we can consider H_2^+ as being formed from a hydrogen atom, H, and a hydrogen ion, H^+. To represent the electron in the atom, we place a half-headed arrow in the 1s box. The 1s box for the ion is left empty because H^+ has no electrons. H_2^+ has one electron and this goes into the box $(1s_A + 1s_B)$.

H_2^+ is easy to deal with because it has only one electron, but it is not a very common molecule. Can we use the idea of combining atomic orbitals when we come to study molecules with more than one electron? Let us start with H_2, dihydrogen, a more familiar molecule.

H_2 has two electrons and we can use a building-up process similar to that used to determine electronic configuration for atoms. The orbital $(1s_A + 1s_B)$ can take two electrons of opposite spin. If we draw a molecular orbital diagram for H_2, we put one electron in each 1s box and two paired electrons in the $(1s_A + 1s_B)$ box, as shown in Figure 4.4. The two hydrogen atoms in H_2 are bonded together by a pair of electrons in a bonding orbital. We have an electron-pair bond.

We should point out here that the bonding orbital $(1s_A + 1s_B)$ will not be exactly the same for H_2^+ and H_2. As with atoms, when we have more than one electron we have to allow for the other electron(s) when we calculate an orbital.

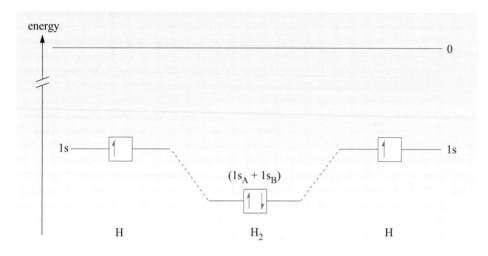

Figure 4.4
Energy-level diagram for H_2.

The zero level in diagrams such as Figure 4.4 refers to free nuclei and free electrons, that is for H_2, two protons and two electrons. The energy for the 1s electron in a hydrogen atom corresponds to the process:

$$H^+(g) + e^-(g) = H(g) \qquad (4.1)$$

This is the reverse of ionization. To ionize an atom requires the input of energy, thus in the process shown energy is released.

⬤ What will be the sign of the energy for the hydrogen 1s level?

⬤ Negative. Energy is released when the electron combines with the proton to form a hydrogen atom and so the electron in the hydrogen atom is at a lower energy than a free electron.

The energy of the electron in the bonding orbital $(1s_A + 1s_B)$ is also negative and more negative than that of the electron in the hydrogen 1s level. In the hydrogen molecule, there are thus two electrons with lower energy than they would have in hydrogen atoms. The energy of the electrons is thus lower for the hydrogen molecule than for the hydrogen atoms, but does this show that hydrogen is more stable as a diatomic molecule than as atoms?

For this to be so, the Gibbs function change,[*] $\Delta G_m^\ominus$, for Equation 4.2

$$2H(g) = H_2(g) \qquad (4.2)$$

* Thermodynamic quantities are discussed in *Metals and Chemical Change*.[1]

must be negative. The energy of the electrons contributes to $\Delta H_m^\ominus$, but there is another term that also contributes, and that is the electrostatic repulsion between the nuclei. In H_2, the nuclei are fixed 74 pm apart. For the two separate hydrogen atoms (2H) the nuclei are a very long way apart and this contribution is small enough to be disregarded. This term increases the energy calculated for the electrons in molecules. In most cases, however, it is not the determining factor in deciding on the stability of molecules.

The other term we have neglected is the entropy term. When we are comparing molecules with free atoms at room temperature, it turns out that the enthalpy term, $\Delta H_{\mathrm{m}}^{\ominus}$ is the more important.

We can therefore make the following generalization:

> Molecular orbital theory predicts that if the total energy of the electrons in a molecule is lower (more negative) than the total electron energy of its constituent atoms, then that molecule will be stable with respect to the free atoms.

4.1.1 Bonding and antibonding orbitals

So far we have only one molecular wavefunction, but for molecules with more than two electrons we are obviously going to need more. Let's start by having another look at how we combined the 1s orbitals to form the molecular orbital for H_2^+ and H_2. We considered the 1s orbital as drawn in Figure 4.5a and we combined it with another similar one. But what about the 1s orbital in Figure 4.5b? This is exactly the same as that in Figure 4.5a except for the phase.

Do we know what phase a 1s orbital has? In an atom, any property we want to calculate, such as the energy or the electron distribution, involves the electron density, the square of the wavefunction, and this tells us nothing about the *sign* of the wavefunction. So no observable property of an isolated atom will tell us the phase of the 1s orbital. When we want to combine two 1s orbitals, however, it matters very much whether the two have the same sign or not. In the previous Section we took two 1s orbitals with the same sign, but suppose we took two 1s orbitals of opposite sign.

● Figure 4.6 shows a diatomic molecule in which 1s orbitals of opposite phase are combining to form a molecular orbital. What can you say about the probability of an electron being found between the two nuclei?

● Between the nuclei, the positive and negative 1s orbitals overlap. When we add together wavefunctions of opposite phase, they cancel each other out. There is therefore very little chance of finding the electron in the region between the nuclei.

Midway between the nuclei, there is no chance of finding the electron because there is a **nodal plane** at right-angles to the molecular axis, where the electron density is zero.

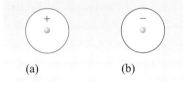

Figure 4.5
1s orbitals with (a) positive phase and (b) negative phase.

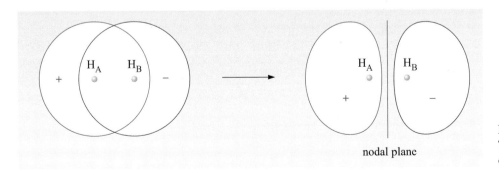

nodal plane

Figure 4.6
The combination of two 1s orbitals of opposite phase.

We shall label this second orbital $(1s_A - 1s_B)$. (It does not matter whether we take the combination $(1s_A - 1s_B)$ or the combination $(1s_B - 1s_A)$; both will have the same observable properties.)

An electron in an $(1s_A - 1s_B)$ orbital has even less chance of being attracted to both nuclei than one in a 1s orbital on one of the atoms. This orbital is therefore of higher energy than the 1s atomic orbital.

The orbital $(1s_A + 1s_B)$ is a bonding orbital. An electron in a bonding orbital is very likely to be found between the nuclei, and thus draws the nuclei together or, in other words, bonds them together. Its energy will be less than that of the constituent atomic orbitals.

The orbital $(1s_A - 1s_B)$ is an **antibonding orbital**. An electron in this orbital is less good at drawing the two nuclei together than an electron in an atomic orbital would be. In fact it tends to keep the nuclei apart. An electron in such an orbital has a higher energy than an electron in one of the atomic orbitals from which it is made.

Figure 4.7 shows a **molecular orbital energy-level diagram** for H_2. As with the energy-level diagram for H_2^+, energy increases from the bottom to the top of this diagram. Hence, the bonding orbital $(1s_A + 1s_B)$ is placed below the atomic orbitals, and the antibonding orbital $(1s_A - 1s_B)$ above.

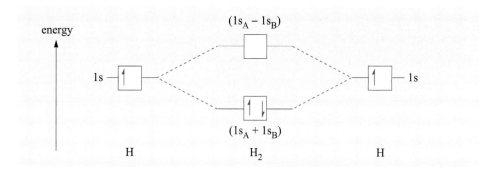

Figure 4.7
Molecular orbital energy-level diagram for H_2, showing the antibonding orbital.

QUESTION 4.1

The molecule dihelium, He_2, would have four electrons. Use the orbital energy-level diagram for H_2 (Figure 4.7), and the rules that we used to assign electrons to orbitals in atoms, to determine the orbital energy-level diagram for this molecule. Represent the electrons as arrows in boxes. How will the energy of this molecule compare with that of two He atoms? He_2 has never been observed; does your energy-level diagram suggest a reason why this should be so?

4.1.2 Summary of Section 4.1.1

1   An electron in a bonding orbital holds the nuclei together and is of lower energy than if it were in one of the atomic orbitals from which the bonding orbital is made. Bonding orbitals are in-phase combinations of atomic orbitals.

2   An electron in an antibonding orbital is of higher energy than if it were in a constituent atomic orbital. Antibonding orbitals are out-of-phase combinations of atomic orbitals.

4.1.3 Molecular orbitals from s orbitals

We are now going to see how we can make molecular orbitals from atomic orbitals other than the 1s by applying the same principles that we used to construct the $(1s_A + 1s_B)$ and $(1s_A - 1s_B)$ molecular orbitals. Let's start with the 2s orbitals. A 2s orbital is spherically symmetrical, like a 1s orbital, and we can combine it with another 2s orbital in exactly the same way as we combined two 1s orbitals.

⬤ What molecular orbitals do you think we could make from two 2s orbitals, one on each of the two nuclei A and B?

⬤ We could combine two 2s orbitals of the same phase to make $(2s_A + 2s_B)$, or we could combine two 2s orbitals of opposite phase to make $(2s_A - 2s_B)$.

But will 1s and 2s orbitals combine? How do the energies of molecular orbitals formed from different atomic orbitals compare? To answer these questions we first need to note the following important point. *In general to form molecular orbitals, the atomic orbitals that combine must be of similar energy.*

The energy difference between the 1s and 2s orbitals is large, about $4\,800\,\text{kJ}\,\text{mol}^{-1}$ for Li, and so the 2s orbitals combine together but separately from the two 1s orbitals.

Consider the gas-phase molecule Li_2. Lithium, Li, has the electronic configuration $1s^2 2s^1$, and Li_2 will therefore have six electrons. We combine two 1s atomic orbitals to make two molecular orbitals $(1s_A + 1s_B)$ and $(1s_A - 1s_B)$. The two 2s atomic orbitals combine to make two more. What energies will these orbitals have? The $(1s_A + 1s_B)$ molecular orbital will be of lower energy than a lithium 1s atomic orbital. The $(1s_A - 1s_B)$ molecular orbital will be of higher energy than the 1s atomic orbital. Similarly, the $(2s_A + 2s_B)$ molecular orbital will be of lower energy and the $(2s_A - 2s_B)$ molecular orbital of higher energy than the 2s orbital in the lithium atom. Because of the large energy difference between the 1s and 2s orbitals in Li the $(2s_A + 2s_B)$ orbital will be higher in energy than the $(1s_A - 1s_B)$ orbital. Each orbital can contain two electrons of opposite spin, so we can now feed our electrons into the Li_2 orbitals, using the same rules as we employed in the case of atoms. The first two go into $(1s_A + 1s_B)$, the next two into $(1s_A - 1s_B)$ and the last two into $(2s_A + 2s_B)$. The orbital energy-level diagram is shown in Figure 4.8.

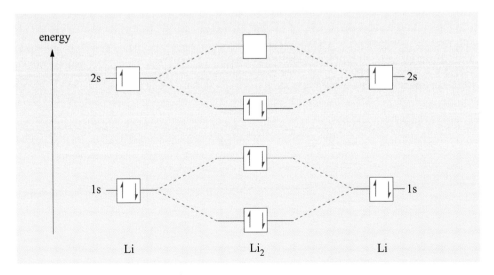

Figure 4.8
Orbital energy-level diagram for $Li_2(g)$.

● Would you expect $Li_2(g)$ to be stable with respect to two Li atoms?

● Yes, because the electrons in the $(2s_A + 2s_B)$ orbital are of lower energy than they would be in a 2s orbital. The energy contributions of the $(1s_A + 1s_B)$ and $(1s_A - 1s_B)$ electrons effectively cancel out relative to the 1s atomic orbital.

We have predicted that Li_2 is stable with respect to lithium atoms, so we might expect to be able to observe dilithium molecules. At room temperature, lithium is a metal, but if we heat lithium up sufficiently, we do indeed find Li_2 molecules present in the vapour. From Figure 4.8 we can only draw conclusions about the relative stabilities of $Li_2(g)$ and $Li(g)$. When we say that a particular molecule is stable with respect to the atoms from which it is made, you should bear in mind that it may not be stable with respect to other molecules or to solid forms of the element or compound.

Before we go on to discuss other molecules, there are some definitions and labels that need to be introduced.

Until now we have labelled our molecular orbitals by using the labels of the atomic orbitals used to make them. This is fine for simple molecules but can become rather cumbersome for larger ones. A standard labelling for molecular orbitals has therefore been developed. This depends on the behaviour of orbitals when acted on by symmetry operations. Symmetry operations will be formally introduced later. Here we need to discuss just two.

One is rotation about the bond. The orbital $(1s_A + 1s_B)$ is oval when viewed facing the bond as in Figure 4.2, but viewed end-on along the bond is circular (Figure 4.9). If we rotate the $(1s_A + 1s_B)$ orbital about the bond, we find that whatever angle we rotate it through it always looks the same. Orbitals in diatomic and other linear molecules that behave like this are called σ **orbitals** (sigma, the Greek equivalent of s).

The orbital $(1s_A - 1s_B)$ will also give an identical orbital if we rotate it about the bond. It too is a σ orbital. To distinguish the two, we have to consider another symmetry operation. If you look at Figure 4.10, where we reproduce $(1s_A + 1s_B)$ and $(1s_A - 1s_B)$, you can see that whereas $(1s_A + 1s_B)$ retains the same sign throughout, one side of $(1s_A - 1s_B)$ is of opposite sign to the other. We can describe this difference by using the idea of a centre of symmetry. A **centre of symmetry** is a point in the centre of a molecule such that if a line is drawn from any atom in the molecule to the centre and then continued the same distance the other side then an identical atom is reached. Thus the midpoint of the H—H bond in the hydrogen molecule is a centre of symmetry. The process of going from any point in space through the centre of symmetry to an equidistant point is called **inversion through the centre of symmetry**.

Figure 4.9
The orbital $(1s_A + 1s_B)$ in H_2 viewed along the H—H bond.

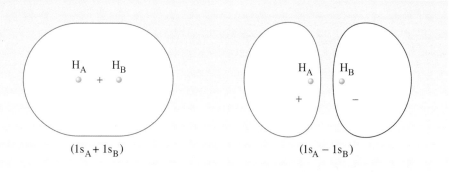

$(1s_A + 1s_B)$ $(1s_A - 1s_B)$

Figure 4.10
$(1s_A + 1s_B)$ and $(1s_A - 1s_B)$ orbitals in H_2.

In Figure 4.11 we have inverted the orbital $(1s_A + 1s_B)$ through the centre of symmetry of the hydrogen molecule.

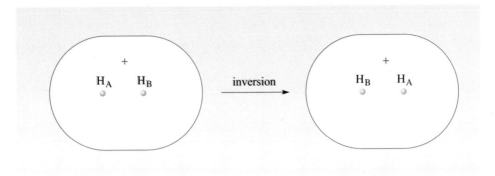

Figure 4.11
Inversion of $(1s_A + 1s_B)$ through the centre of symmetry.

● How does the orbital $(1s_A + 1s_B)$ behave when it is inverted through the centre of the molecule?

● It is turned into an identical orbital.

● How does the orbital $(1s_A - 1s_B)$ behave when it is inverted through the centre of the molecule?

● Inversion of this orbital through the centre of the molecule gives *minus* the original orbital (see Figure 4.12).

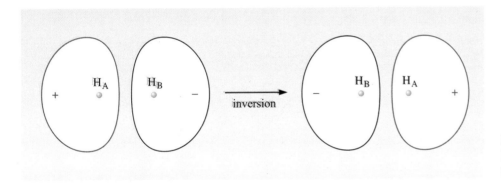

Figure 4.12
Inversion of the orbital $(1s_A - 1s_B)$ through the centre of the molecule.

Orbitals such as $(1s_A + 1s_B)$, which give an identical orbital if inverted through the centre of the molecule are distinguished by the subscript g.[*]

Orbitals such as $(1s_A - 1s_B)$, which give minus the original orbital when inverted through the centre are given the subscript u.[*] So $(1s_A + 1s_B)$ is a σ_g orbital and $(1s_A - 1s_B)$ is a σ_u orbital.

● How would you label $(2s_A + 2s_B)$ and $(2s_A - 2s_B)$?

● $(2s_A + 2s_B)$ is σ_g, and $(2s_A - 2s_B)$ is σ_u.

We now have two σ_g orbitals and two σ_u orbitals. If we need to distinguish them we can do so by labelling them with the numbers 1, 2, 3, ..., starting with the lowest energy orbital with a particular symmetry label. So $(1s_A + 1s_B)$ becomes $1\sigma_g$ and

[*] g stands for the German *gerade*, meaning even; u stands for *ungerade*, meaning odd.

45

$(2s_A + 2s_B)$ becomes $2\sigma_g$. These numbers tell us nothing more than the order of energies of the molecular orbitals; *they are not quantum numbers.*

Alternatively, we can label them with the atomic orbitals they are formed from, so that $(1s_A + 1s_B)$ becomes $1s\sigma_g$ and $(2s_A + 2s_B)$ becomes $2s\sigma_g$. Figure 4.13 shows the orbital energy-level diagram for $Li_2(g)$ using these symmetry labels.

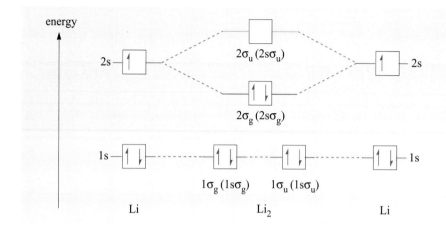

Figure 4.13
Orbital energy-level diagram for $Li_2(g)$, showing symmetry labels.

You have probably noticed there is one further difference between Figures 4.8 and 4.13. In Figure 4.13 the $1\sigma_g$ and $1\sigma_u$ orbitals are drawn as having the same energy as the 1s orbitals. Why should this be so?

- When we discussed atoms with more than one electron, we pointed out that although the orbitals in these atoms are similar to those of the hydrogen atom, they were not identical with hydrogen atom orbitals. Can you remember what properties we mentioned in particular as varying from atom to atom?

- The size and energy of the orbitals.

- How do the sizes of the 1s orbitals in Li and H compare?

- The 1s orbital in Li is smaller (Figure 3.17).

- The distance between the hydrogen atoms in H_2 is 74 pm. The Li—Li bond length in Li_2 is 270 pm. How does the overlap of two 1s orbitals in H_2 compare with the overlap in Li_2?

- The Li 1s orbitals are smaller than those of hydrogen and further apart. The overlap of the H 1s orbitals is much greater.

This is illustrated in Figure 4.14. The difference in energy between the $1\sigma_g$ and 1s orbitals is a lot smaller in Li_2 than in H_2, so that although the energies of $1\sigma_g$ and $1\sigma_u$ differ from that of a 1s orbital, the differences can be neglected relative to those between $2\sigma_g$ and $2\sigma_u$ and the 2s orbitals.

We find that, in general, there is very little overlap between atomic orbitals lower in energy than those of the valence-shell electrons. These orbitals are often left out of the molecular orbital energy-level diagrams. This makes it easier to concentrate on the chemically more interesting valence orbitals.

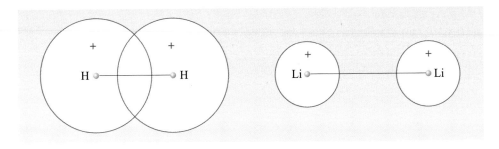

Figure 4.14
Overlap of 1s orbitals in dihydrogen and dilithium.

Now try the following question to see if you have understood how to form molecular orbitals from s orbitals. We then go on to look at molecules where we need to consider p orbitals.

QUESTION 4.2

Draw an orbital energy-level diagram for diberyllium, Be_2. Label the molecular orbitals with their symmetry labels and represent the electrons as arrows in boxes. The spectrum of Be_2 has been observed at low temperature. Would you have predicted the existence of $Be_2(g)$ from your diagram? What is the normal form of beryllium at room temperature and atmospheric pressure?

4.1.4 Molecular orbitals from p orbitals

When we move beyond beryllium to more chemically interesting elements such as carbon, nitrogen and oxygen, the occupied valence orbitals include the 2p.

So far, we have considered only molecular orbitals made by combining s atomic orbitals. We shall now show how we can combine 2p atomic orbitals. There are three 2p orbitals, $2p_x$, $2p_y$ and $2p_z$, directed along three perpendicular axes (Figure 4.15). Can we combine any one of these with any other one, or are there restrictions?

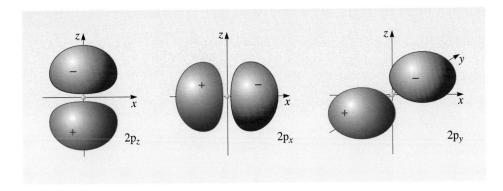

Figure 4.15
$2p_x$, $2p_y$ and $2p_z$ orbitals.

First we have to decide which direction is x, which is y and which is z, for the molecule. or diatomic molecules it is conventional to take the z direction along the molecular axis.

Figure 4.16a (overleaf) shows a $2p_z$ orbital on nucleus A in the molecule N_2. A $2p_z$ orbital on nucleus B is shown in Figure 4.16b. Note that a $2p_z$ orbital is defined as having the positive lobe of the wavefunction in the positive z direction. This is merely a convention, because we do not know which way round the positive and negative lobes are. The convention is necessary to allow us to say whether a

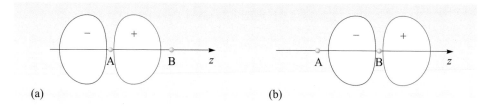

Figure 4.16
$2p_z$ orbitals on (a) nucleus A and (b) nucleus B in dinitrogen.

particular combination of 2p orbitals will overlap to increase the wavefunction between the nuclei or cancel it out.

○ Can you say anything about the combination of $2p_z$ orbitals as they are drawn in Figure 4.16?

○ If you look at this Figure you will see that, between the nuclei, one orbital has a positive lobe and the other a negative lobe. They will therefore cancel each other out where they overlap and there will be little chance of finding the electron between the nuclei.

The molecular orbital ($2p_{zA} + 2p_{zB}$) will therefore be an antibonding orbital of higher energy than a $2p_z$ atomic orbital. This is shown in Figure 4.17 where you can see that the electron density is reduced between the nuclei and increased outside them. Can we give this orbital a symmetry label?

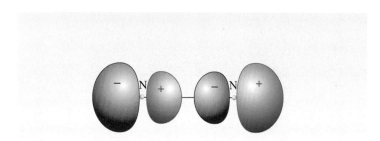

Figure 4.17
The orbital ($2p_{zA} + 2p_{zB}$).

○ Is the orbital ($2p_{zA} + 2p_{zB}$) a σ orbital?

○ Yes. In a linear molecule the label σ applies to any orbital that is unchanged by rotation about the molecular axis through any angle.

○ If we invert the orbital through the centre of the molecule, do we obtain an identical orbital?

○ No. We produce minus the original orbital; ($2p_{zA} + 2p_{zB}$) must therefore take the subscript u.

○ What combination of $2p_z$ orbitals will give us a bonding orbital?

○ We can make a bonding combination by reversing the signs on one of the $2p_z$ atomic orbitals. This gives us the molecular orbital ($2p_{zA} - 2p_{zB}$) shown in Figure 4.18.

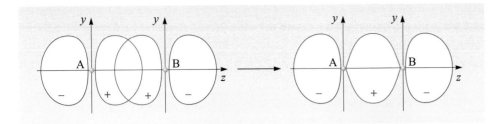

Figure 4.18
A bonding molecular orbital made by combining two $2p_z$ atomic orbitals.

This orbital will be labelled $3\sigma_g$ (being higher in energy than the σ_g orbitals formed by combining 1s and 2s orbitals) or $2p\sigma_g$. Although it looks different from σ_g orbitals made by combining s orbitals, it is still labelled σ_g since it behaves in the same way when rotated about the molecular axis or inverted through the centre of symmetry.

Now let's combine two $2p_y$ orbitals. The lobes of these orbitals (like those of the $2p_x$ orbitals but unlike those of $2p_z$), are perpendicular to the molecular axis. Figure 4.19 shows the combination $(2p_{yA} + 2p_{yB})$. In this combination the amplitude of the wavefunction is increased between the nuclei but only above and below the molecular axis.

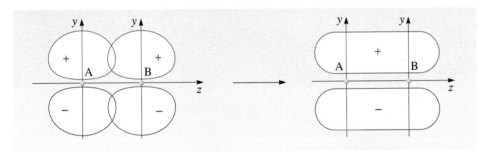

Figure 4.19
The bonding orbital $(2p_{yA} + 2p_{yB})$.

The probability of finding the electron on the molecular axis in this orbital is zero; that is, the molecular axis lies in a nodal plane. The high probability just above and below the axis, however, does serve to draw the nuclei together, and so this orbital is bonding. It is a different sort of bonding orbital from the σ_g orbitals we have already met. $(2p_{yA} + 2p_{yB})$ produces an identical orbital only if we rotate it about the molecular axis through a complete revolution. If we rotate it through half a revolution about this axis, we produce an orbital with the signs reversed. Orbitals like $(2p_{yA} + 2p_{yB})$ are called **π orbitals** (pi, the Greek equivalent of p). You will see later that π orbitals are important for unsaturated organic molecules such as ethene and benzene.

- Should $(2p_{yA} + 2p_{yB})$ be given the subscript g or u?

- Inversion through the centre produces minus the original orbital, so the correct subscript is u.

So $(2p_{yA} + 2p_{yB})$ is a bonding orbital, to which we give the symbol π_u.

- What would the antibonding orbital formed from $2p_y$ orbitals be?

- $(2p_{yA} - 2p_{yB})$.

$(2p_{yA} - 2p_{yB})$ has the same behaviour as $(2p_{yA} + 2p_{yB})$ when rotated about the molecular axis but does not change sign when inverted through the centre of symmetry and so has the symbol π_g. This is shown in Figure 4.20.

We can make, therefore, two molecular orbitals, one bonding (π_u) and one antibonding (π_g), by combining $2p_y$ atomic orbitals. Can we make two molecular orbitals by combining $2p_x$ orbitals? Yes. The $2p_x$ orbitals can form two combinations. These will be identical with those of $2p_y$, except that they will be at right angles to them.

Atomic orbitals that have the same energy but differ only in direction (for example $2p_x$, $2p_y$ and $2p_z$) are called **degenerate**. Similarly, the molecular orbitals $(2p_{xA} + 2p_{xB})$ and $(2p_{yA} + 2p_{yB})$ will be degenerate. The two degenerate bonding orbitals are given the symbol $1\pi_u$ or $2p\pi_u$. The two degenerate antibonding orbitals are given the symbol $1\pi_g$ or $2p\pi_g$.

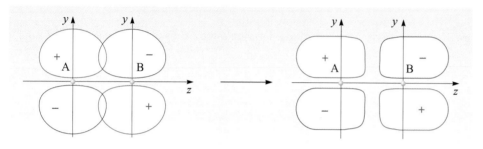

Figure 4.20
An antibonding π orbital formed by combination of two $2p_y$ atomic orbitals.

σ orbitals, like s orbitals, are non-degenerate, whereas there are two π orbitals of the same energy.

Let us try, then, to construct an orbital energy-level diagram for a molecule that has molecular orbitals made from 2p orbitals — the dinitrogen molecule, N_2. The nitrogen atom has seven electrons, so N_2 will have fourteen. We always have to start with the molecular orbital of lowest energy, so we begin with those formed from the nitrogen 1s orbitals, the $1\sigma_g$ ($1s\sigma_g$) and $1\sigma_u$ ($1s\sigma_u$). Then we have the molecular orbitals formed from the 2s atomic orbitals, the $2\sigma_g$ ($2s\sigma_g$) and the $2\sigma_u$ ($2s\sigma_u$).

● How many molecular orbitals can we make from the 2p atomic orbitals on nitrogen?

● Six. $3\sigma_g$ ($2p\sigma_g$) (bonding), $3\sigma_u$ ($2p\sigma_u$) (antibonding), two $1\pi_u$ ($2p\pi_u$) (the two bonding orbitals from $2p_x$ and $2p_y$) and two $1\pi_g$ ($2p\pi_g$) (the two antibonding orbitals from $2p_x$ and $2p_y$).

$3\sigma_g$ ($2p\sigma_g$) and $1\pi_u$ ($2p\pi_u$) were the bonding combinations, so they must be of lower energy than the 2p orbital. But which one will have the higher energy? Look at the $3\sigma_g$ ($2p\sigma_g$) (f) and one of the $1\pi_u$ ($2p\pi_u$) (e) orbitals in Figure 4.21. In the $2p\sigma_g$ orbital, nuclei are held together by electron density directly between them, whereas in $2p\pi_u$ the electron density is holding them together above and below the molecular axis. It seems reasonable that an electron in $2p\sigma_g$ would be better at drawing the nuclei together than one in $2p\pi_u$, and calculations show that for orbitals made purely from p orbitals this is so. $2p\sigma_g$ is thus lower in energy than $2p\pi_u$.

Figure 4.22 (p. 52) shows the energy levels for N_2 that we have built up. At the bottom are $1s\sigma_g$ and $1s\sigma_u$, shown at the same energy since the overlap of 1s orbitals is so small. Next we have the bonding $2s\sigma_g$ below and the antibonding $2s\sigma_u$ above the 2s. Then we have $2p\sigma_g$, the two degenerate $2p\pi_u$ bonding orbitals and at the top the $2p\pi_g$ and $2p\sigma_u$ antibonding orbitals.

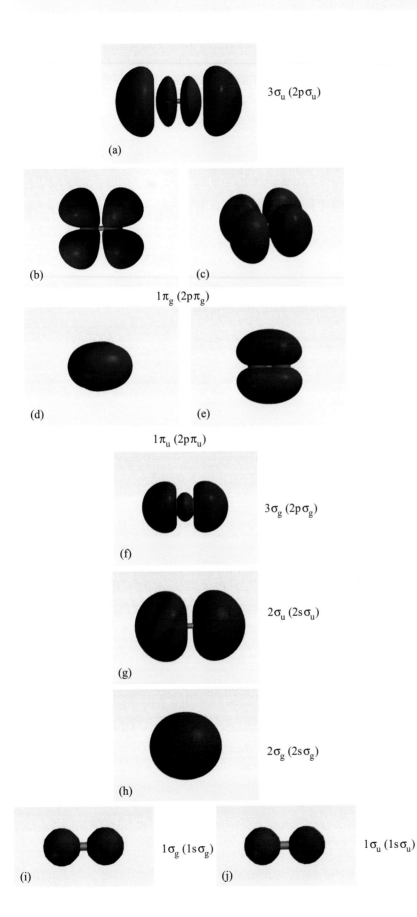

$3\sigma_u$ $(2p\sigma_u)$

(a)

(b)

(c)

$1\pi_g$ $(2p\pi_g)$

(d)

(e)

$1\pi_u$ $(2p\pi_u)$

$3\sigma_g$ $(2p\sigma_g)$

(f)

$2\sigma_u$ $(2s\sigma_u)$

(g)

$2\sigma_g$ $(2s\sigma_g)$

(h)

$1\sigma_g$ $(1s\sigma_g)$

(i)

$1\sigma_u$ $(1s\sigma_u)$

(j)

Figure 4.21
Molecular orbitals for dinitrogen.

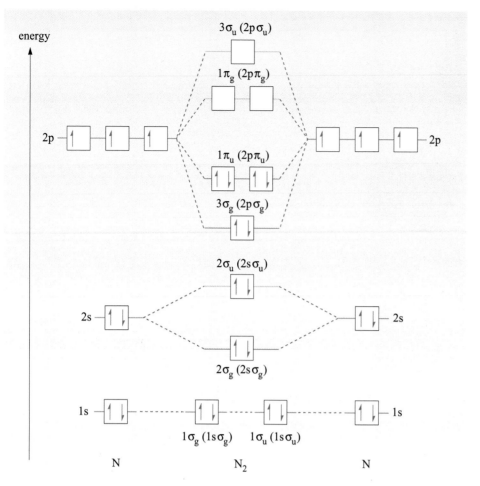

Figure 4.22
Orbital energy-level diagram for dinitrogen.

As the $2p\sigma_g$ is more bonding than $2p\pi_u$ because of the greater overlap, the $2p\sigma_u$ will be more antibonding than the $2p\pi_g$ and hence will be higher in energy.

Now we have the order of the orbitals, we have to fill in the fourteen electrons. We start from the lowest-energy orbital. Two electrons with paired spin will go into $1s\sigma_g$ and two into $1s\sigma_u$. Then two go into $2s\sigma_g$ and two into $2s\sigma_u$. $2p\sigma_g$ is lower in energy than $2p\pi_u$, and so we fill the $2p\sigma_g$ first and then the $2p\pi_u$. The $2p\pi_u$ can take two electrons with opposite spin, but there are two $2p\pi_u$ orbitals and so these take a total of four electrons. This disposes of all the remaining electrons; $2p\pi_g$ and $2p\sigma_u$ remain unoccupied. The filling of electrons is shown in Figure 4.22.

🔵 How many electrons in bonding orbitals are there in N_2?

🔵 Ten: those in $1s\sigma_g$, $2s\sigma_g$, $2p\sigma_g$ and $2p\pi_u$.

🔵 How many electrons are there in antibonding orbitals?

🔵 Four: those in $1s\sigma_u$ and $2s\sigma_u$.

Dinitrogen has an excess of six electrons in bonding orbitals over those in antibonding orbitals. The structural formula of dinitrogen, $N\equiv N$, shows the two atoms joined by three pairs of electrons in covalent bonds. In molecular orbital theory the number of bonds between two atoms is determined by the excess number of pairs of electrons in bonding orbitals over those in antibonding orbitals.

We can define a quantity called the **bond order**, which is equal to the number of pairs of electrons in bonding orbitals minus the number of pairs in antibonding orbitals. An unpaired electron counts as half a pair and thus contributes $+\frac{1}{2}$ if it is in a bonding orbital or $-\frac{1}{2}$ if it is in an antibonding orbital. In general, we find that the bond order corresponds to the number of electron pairs covalently bonding the atoms together in the Lewis structure.

⬤ What is the bond order of dinitrogen?

⬤ Three. There are five pairs of electrons in bonding orbitals and two pairs in antibonding orbitals.

In the next Section we are going to look at some more diatomic molecules.

4.1.5 Summary of Sections 4.1.3 and 4.1.4

Orbital energy-level diagrams for homonuclear diatomic molecules with s and p valence electrons can be constructed using the following guidelines:

1   Atomic orbitals must be of similar energy in order to combine to make molecular orbitals.

2   n atomic orbitals combine to form n molecular orbitals. Thus from two s orbitals, one on each atom, two molecular orbitals can be made. From six 2p atomic orbitals (three on each atom) we can make six molecular orbitals.

3   Each pair of 2p orbitals will combine to form two molecular orbitals, one bonding and one antibonding.

4   The z axis is defined to be along the bond.

5   The two orbitals formed from the $2p_z$ atomic orbitals are unchanged by rotation through any angle about the molecular axis and are therefore σ orbitals.

6   The $2p_x$ and $2p_y$ orbitals form π orbitals. The sign of these orbitals is reversed when they are rotated through half a revolution about the molecular axis.

7   σ bonding orbitals are unchanged by inversion through the centre of the molecule and are therefore labelled σ_g. π bonding orbitals are changed into minus themselves and are labelled π_u.

8   σ antibonding orbitals are labelled σ_u and π antibonding orbitals are labelled π_g.

9   σ orbitals are non-degenerate; π orbitals are doubly degenerate (that is, there are always two of the same energy).

10   Among the orbitals derived from the 2p atomic orbitals, the σ bonding orbital is of lower energy than the π bonding orbitals, and consequently the π antibonding orbital is of lower energy the σ antibonding orbital.

11   The bond order can be defined as the number of pairs of electrons in bonding orbitals minus the number of pairs of electrons in antibonding orbitals.

QUESTION 4.3

What are the bond orders of the following molecules: (a) H_2; (b) He_2; (c) Li_2?

4.2 Homonuclear molecules of the first and second rows of the Periodic Table

A **homonuclear diatomic molecule** is one in which both nuclei are the same, for example H_2 and N_2. In the first row of the Periodic Table, H_2 is the only example. From the second row we have N_2, O_2 and F_2, which are stable under normal conditions of temperature and pressure. We looked at N_2 in the previous Section. Here we shall consider the molecular orbital description of O_2, and use it as an example of how we can use the theory to explain and/or predict properties of molecules.

As a further example we shall then take C_2. This is not a stable molecule under normal conditions, but it has been detected in such places as flames and the heads of comets.

Oxygen has the electronic configuration $1s^2 2s^2 2p^4$, and so the dioxygen molecule, O_2, has sixteen electrons. Four of these will go into the two σ orbitals formed from 1s orbitals, and four more into the two σ orbitals formed from 2s orbitals. This leaves eight electrons to go into molecular orbitals formed by combining 2p orbitals. The orbitals available will be similar to those for N_2 but will have different energies. So we can put two electrons into the σ bonding orbital $3\sigma_g$ ($2p\sigma_g$) and two into each of the π bonding orbitals $1\pi_u$ ($2p\pi_u$). This disposes of six of the eight electrons. What do we do with the last two?

○ Would you expect these two to go into the π antibonding orbital, $2p\pi_g$ ($1\pi_g$), or the σ antibonding orbital, $2p\sigma_u$ ($3\sigma_u$)?

○ They will go into whichever has the lower energy. When we form a bonding and an antibonding orbital from two atomic orbitals, the antibonding orbital is raised in energy by about the same amount that the bonding orbital is lowered in energy. Since the $2p\sigma_g$ bonding orbital is lower in energy than the $2p\pi_u$ bonding orbitals, the $2p\sigma_u$ antibonding orbital will be higher in energy than the $2p\pi_g$ antibonding orbitals. The electrons will therefore go into the $2p\pi_g$ ($1\pi_g$) orbitals. The orbital energy-level diagram for O_2 is shown in Figure 4.23.

○ In Figure 4.23 we have shown the two electrons in the $1\pi_g$ orbitals with parallel spins. Why have we done this?

○ Electrons in molecules, as well as those in atoms, obey Hund's rule (Section 3.1.1), and since the two $1\pi_g$ orbitals are degenerate the two electrons will go one into each $1\pi_g$ orbital with parallel spins.

If we place liquid oxygen (which consists of O_2 molecules) near a strong magnet, it is drawn into the magnetic field (Figure 4.24). Liquid nitrogen, on the other hand, is hardly affected and is, if anything, slightly repelled from the field. Dioxygen is said to be **paramagnetic**, whereas dinitrogen is said to be **diamagnetic**. Atoms or molecules with unpaired spins have a magnetic moment and are attracted to an applied magnetic field. By predicting that O_2 has two unpaired electrons, molecular orbital theory can account for the behaviour of dioxygen in a magnetic field. It also predicts that dinitrogen will be diamagnetic because all the electrons are paired.

The explanation of the paramagnetism of dioxygen was one of the early triumphs of molecular orbital theory, since the property could not be explained in terms of earlier theories.

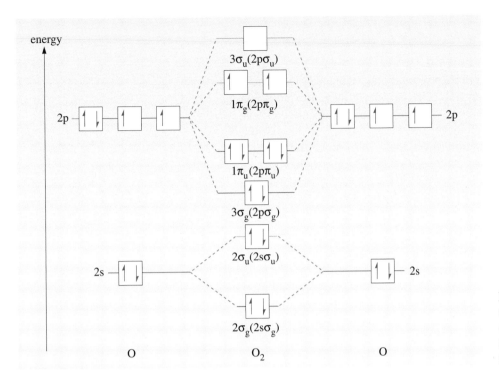

Figure 4.23
Orbital energy-level diagram for dioxygen. 1s orbitals and the molecular orbitals formed from them are omitted for clarity.

Figure 4.24
A stream of liquid oxygen held between the poles of a magnet.

What is the bond order in O_2?

Two: O_2 has three pairs of electrons in 2p bonding orbitals and two half-pairs in antibonding orbitals. The orbitals formed from 1s and 2s are completely filled, and we can neglect them as they will not contribute to the bond order.

As with N_2, the bond order is equal to the number of covalent bonds predicted by the Lewis structure. Are the different bond orders of O_2 (two) and N_2 (three) reflected in their properties?

We would expect that the more electrons there are holding two nuclei together, the harder it will be to force them apart and the closer they will be at equilibrium. Let's then compare the bond dissociation energies and bond lengths of N_2 and O_2. The dissociation energies, that is $\Delta H_m^{\ominus}$ for the processes $N_2(g) \longrightarrow 2N(g)$ and $O_2(g) \longrightarrow 2O(g)$, are 945 kJ mol^{-1} for dinitrogen and 498 kJ mol^{-1} for dioxygen. It thus requires more energy to split the dinitrogen molecule into two nitrogen atoms than to obtain two oxygen atoms from O_2. This is what we would expect from the bond orders. Similarly, the bond length of N_2 (110 pm) is less than that of O_2 (121 pm).

When comparing bond lengths and dissociation energies in this way, you should be careful to choose molecules with similar molecular orbitals. In this case we are comparing two molecules that use the same atomic orbitals (2s and 2p) to form their molecular orbitals. We can thus compare N_2 with O_2 and F_2, but not with H_2 (1s) or Cl_2 (3s and 3p), if we are interested in the effect of bond order. For example, H_2 and F_2 both have bond orders of 1 but very different bond lengths (74 pm and 141 pm, respectively) because the 2p electrons are, on average, further from the F nucleus than the 1s electron is from the H nucleus. The dissociation energies are 436 kJ mol^{-1} and 158 kJ mol^{-1}, respectively. Thus the bond in H_2 is almost half as long and over twice as strong as that in F_2 even though H_2 and F_2 have the same bond order.

4.2.1 The molecule C_2

Look again at the orbital energy-level diagrams for N_2 and O_2 (Figures 4.22 and 4.23). Which molecular orbitals would you expect to be filled in C_2?

C_2 has twelve electrons, so we would expect to fill the σ_g and σ_u formed from 1s, $2s\sigma_g$, $2s\sigma_u$, and $2p\sigma_g$ orbitals, and have two electrons in the $2p\pi_u$ orbitals.

Spectroscopic observations of C_2 indicate that in the ground state the molecule has no unpaired electrons.

Does this evidence fit the orbital occupancy we predicted above?

No. By Hund's rule the two electrons in the $2p\pi_u$ orbitals would be expected to be unpaired.

So far, we have made our molecular orbitals either entirely from s orbitals or entirely from p orbitals. For the hydrogen atom, however, 2s and 2p orbitals are of the same energy, and for the lighter atoms of the second row of the Periodic Table the difference between the energies of the 2s and 2p orbitals is still quite small. Perhaps, then, we should consider the overlap of 2s and 2p orbitals. As you saw earlier the shape of molecules containing carbon can be successfully modelled using hybrid orbitals that are a mixture of 2s and 2p.

Does the possibility of overlap between 2s and $2p_z$ atomic orbitals affect the orbital energy-level diagram for C_2? If the carbon 2s and 2p atomic orbitals are close enough together in energy, then in constructing the $2\sigma_g$, $2\sigma_u$, $3\sigma_g$ and $3\sigma_u$ molecular orbitals we should consider the 2s and $2p_z$ atomic orbitals together. Let's look at the formation of the $2\sigma_g$ and the $3\sigma_g$ molecular orbitals. We have regarded the $2\sigma_g$ orbitals so far as a combination of 2s orbitals, and the $3\sigma_g$ as a combination of $2p_z$ orbitals. In molecules like C_2, however, $2\sigma_g$ and $3\sigma_g$ will both be formed from a combination of 2s and 2p orbitals. The addition of 2p to $2\sigma_g$ occurs in phase, adding bonding character and will lower its energy. The addition of 2s to $3\sigma_g$, however, occurs out of phase, i.e. in an antibonding manner, and raises the energy of this orbital.

⬤ How does raising the energy of the $3\sigma_g$ orbital affect the orbital energy-level diagram of a molecule such as C_2?

⬤ If the energy of the $3\sigma_g$ orbital is raised sufficiently, it may become of higher energy than the $1\pi_u$.

⬤ If the $3\sigma g$ orbital were higher in energy than the $1\pi_u$, would this explain why C_2 has no unpaired electrons in its ground state?

⬤ Yes. The twelve electrons would then fill the $1\sigma_g$, $1\sigma_u$, $2\sigma_g$, $2\sigma_u$ and $1\pi_u$ orbitals. There would be four electrons in the $1\pi_u$ orbitals, and these would be paired; there would be none in the $3\sigma_g$.

The revised orbital energy-level diagram for C_2 is shown in Figure 4.25. Here we have dropped the alternative labelling, $2p\sigma_g$, etc., since 2s and 2p are both involved in some orbitals, so it is no longer appropriate to label $2\sigma_g$ as $2s\sigma_g$ or $3\sigma g$ as $2p\sigma_g$.

If we look at calculated molecular orbitals, we find that there is always a mixture of atomic orbitals but that as we go across the Periodic Table the amount of 2s in $3\sigma_g$ and $2p_z$ in $2\sigma_g$ diminishes, so that the properties of N_2, O_2 and F_2 can be explained qualitatively by regarding $3\sigma_g$ as entirely composed of $2p_z$ orbitals.

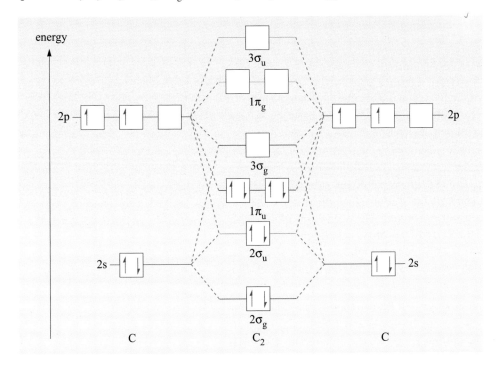

Figure 4.25
Orbital energy-level diagram for C_2.

4.2.2 Summary of Section 4.2

1   Molecular orbitals for homonuclear diatomic molecules can be formed by adding together atomic orbitals on each atom.

2   Molecular orbital occupancies in such diatomic molecules predict bond orders and magnetic properties. High bond orders correspond to large dissociation energies and short bond lengths.

3   If the 2s and 2p atomic orbitals are close in energy, the molecular orbitals $2\sigma_g$ and $3\sigma_g$ are no longer purely 2s or 2p but mixtures of these. This can result in the $3\sigma_g$ orbital being of higher energy than the $1\pi_u$.

QUESTION 4.4

Draw an orbital energy-level diagram for dineon, Ne_2. What is the bond order of this molecule? Would you predict *from your diagram* that this molecule should exist?

QUESTION 4.5

Draw an orbital energy-level diagram for O_2^+. [Use the diagram for O_2 (Figure 4.23) but feed in one fewer electron.] How would you expect (a) the bond length and (b) the dissociation energy of this ion to differ from those of O_2?

QUESTION 4.6

Use the orbital energy-level diagram for N_2 (Figure 4.22) to find the bond orders of N_2^+ and N_2^-. How would you expect the dissociation energies of these ions to compare with each other and with that of N_2?

4.3 Homonuclear diatomic molecules of later rows

Similar diagrams to those for N_2, O_2 and C_2 can be drawn for molecules composed of atoms from later rows. Take, for example, dichlorine, Cl_2. Chlorine has the electronic configuration $1s^2 2s^2 2p^6 3s^2 3p^5$. Here we can regard the 1s, 2s and 2p orbitals as being effectively atomic orbitals, and consider only the valence-shell orbitals 3s and 3p. In chlorine there is a relatively large energy gap between 3s and 3p, and so, as in N_2 and O_2, we concentrate on the p orbitals. $3p_x$, $3p_y$, and $3p_z$ orbitals combine in just the same way as $2p_x$, $2p_y$ and $2p_z$ to form σ_g and π_u bonding orbitals, and σ_u and π_g antibonding orbitals. If we number these orbitals, however, we have to allow for the 2s and 2p. So, the 1s orbitals would provide $1\sigma_g$ and $1\sigma_u$ molecular orbitals (these are labelled as molecular orbitals even though there is little overlap between 1s orbitals on the two atoms). The 2s orbitals would give $2\sigma_g$ and $2\sigma_u$, the 2p would give $3\sigma_g$, $1\pi_u$, $1\pi_g$ and $3\sigma_u$, and the 3s would give $4\sigma_g$ and $4\sigma_u$. The orbitals formed from the 3p orbitals are thus labelled $5\sigma_g$ (or $3p\sigma_g$), $2\pi_u$ ($3p\pi_u$), $2\pi_g$ ($3p\pi_g$) and $5\sigma_u$ (or $3p\sigma_u$). An orbital energy-level diagram for Cl_2 is shown in Figure 4.26.

For Br_2, the 4p orbitals would form a similar set, but the numbering would go even higher as we would have to allow for 1s, 2s, 2p, 3s, 3p, 4s and 3d orbitals. The s and p orbitals will combine as in Cl_2, the 3d orbitals form two σ orbitals, four π orbitals and four orbitals of a new type, δ (delta, the Greek equivalent of d).

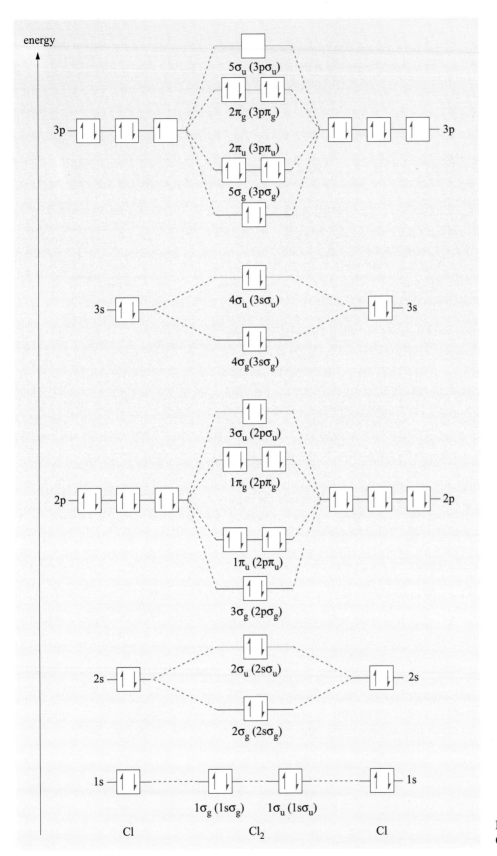

Figure 4.26
Orbital energy-level diagram for Cl_2.

QUESTION 4.7

Although elemental sulfur is a solid under normal conditions of temperature and pressure, S_2 molecules can be observed in sulfur vapour when the solid is heated. Draw an orbital energy-level diagram for S_2, showing the orbitals made from the 3p on S. Predict the bond order of S_2 and state whether or not it will be paramagnetic.

4.4 Heteronuclear diatomic molecules

There are only a few heteronuclear diatomic molecules that are formed from elements of the first and second rows of the Periodic Table and are stable as diatomic molecules in the gas phase at normal temperatures and pressures. These are HF, CO and NO. Others have been observed at high temperatures, in discharge lamps, in flames or in space. Examples are LiH, LiF, OH, BeH, BeO, BF, BH, CH, CN and NH. Some of the molecules in this second list will be stable with respect to the two separate atoms but not at normal temperatures and pressures with respect to other forms of the compound. LiH, LiF and BeO are normally found as ionic solids. The other molecules are unstable with respect to covalent compounds in which the atoms have their normal valencies: H_2O, BeH_2, BF_3, B_2H_6, CH_4, $(CN)_2$ and NH_3.

As an example, let's look at the molecular orbitals of nitric oxide (NO). Nitric oxide at room temperature and atmospheric pressure is a colourless gas. The gas consists mainly of paramagnetic NO molecules, but at lower temperatures nitric oxide condenses to a liquid and then a solid containing dimers, N_2O_2. On contact with air, NO reacts with oxygen gas to form nitrogen dioxide, NO_2. It also reacts readily with halogens, other than iodine, to give nitrosyl halides, NOF, NOCl, and NOBr.

In the late twentieth century, nitric oxide was revealed as a versatile biologically active molecule. For example, it has a significant role in regulating blood pressure. The mechanism by which it does this is not entirely clear but it has been proposed that nitric oxide binds to both the iron and to sulfur in haemoglobin. When nitric oxide bound to sulfur is transferred to receptors, the blood vessels dilate and this leads to a drop in blood pressure. In investigating the mechanism by which NO acts in biological systems, it is important to understand its electronic structure because this can tell us how it might bind to other molecules.

The NO molecule has fifteen electrons. This is an odd number, and hence there must be at least one unpaired electron. The unpaired electron will account for the paramagnetism of the molecule, whichever theory of bonding we use.

To construct molecular orbitals for NO we first have to decide which atomic orbitals to combine. We can combine only atomic orbitals that are close in energy. The orbitals closest in energy are the 2p orbitals on nitrogen and the 2p orbitals on oxygen. The 1s orbitals are too far from each other and from any other orbitals to overlap significantly, so we can forget these. The energies of the 2s and 2p orbitals on each atom are quite well separated, so we shall concentrate only on the 2p, as we did with N_2 and O_2.

The three 2p orbitals from each atom can be combined to give σ bonding and antibonding orbitals and π bonding and antibonding orbitals. In heteronuclear diatomic molecules, these orbitals are simply labelled σ or π and do not have subscripts. The subscripts g and u refer to behaviour under inversion through the centre of symmetry of a molecule. However, because the two atoms are not identical in heteronuclear diatomics, the molecule NO does not have a centre of symmetry, and so we cannot use g and u to describe the orbitals.

For heteronuclear diatomic molecules, both σ bonding and the σ antibonding orbitals have the same symmetry, and thus have to have the same label, σ. Where bonding and antibonding orbitals have the same symmetry label, it is common to distinguish between them by labelling the antibonding orbital with an asterisk. Thus the $2p\sigma$ antibonding orbital may be written $2p\sigma^*$ (two p sigma star).

We shall now construct a partial orbital energy-level diagram for NO using 2p orbitals on both N and O. Look at Figure 4.27. On one side of the diagram we have the N 2p orbitals and on the other side the O 2p orbitals. Because the energies of N 2p and O 2p are not the same, the orbitals are drawn at different levels on the energy scale.

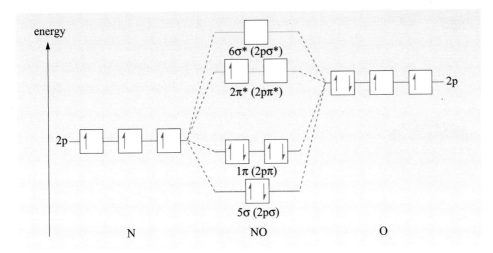

Figure 4.27
Partial orbital energy-level diagram of NO. The molecular orbitals are labelled in the two ways that we introduced for homonuclear diatomic molecules. Note that we reach higher numbers when labelling the σ and π orbitals than we did for N_2 or O_2. This is because we can no longer distinguish g and u, and so have to label all σ and all π orbitals consecutively.

● Why are O 2p and N 2p atomic orbitals not at the same energy?

● Following the earlier example of Li and H, as the nuclear charge increases, the increased electrostatic attraction between the nucleus and an electron will not be offset completely by increased electronic repulsion.

As the oxygen nucleus has the higher nuclear charge, the O 2p orbitals would be expected to be of lower energy than N 2p orbitals. On average, the O 2p electron would be expected to be closer to its nucleus than an N 2p electron to its. In fact the introduction of a 2p electron of opposite spin to the other three in oxygen affects the change in electron repulsion from N to O so that the N 2p level is slightly lower in energy than the O 2p.

From combinations of the 2p orbitals we form σ and π bonding orbitals. These are lower in energy than either of the 2p levels. We can also form antibonding combinations, σ and π, and these lie higher in energy than either 2p level. As with the homonuclear diatomics, the σ bonding orbital is lower in energy than the π bonding orbital and the σ antibonding orbital is higher in energy than the π antibonding orbital.

Apart from the slightly different label, there is another difference between the orbitals of heteronuclear molecules, such as NO, and those of homonuclear molecules, such as N_2. For N_2, combinations such as $(2p_{zA} + 2p_{zB})$ had equal contributions from orbitals on atoms A and B. For NO, contributions of atomic orbitals from N and O are not equal. In general the contribution of an atomic orbital will depend on how close in energy it is to the molecular orbital. The σ and π bonding orbitals in NO are closer in energy to the lower N 2p than to the O 2p and these orbitals contain a greater contribution from N 2p. For $2p\sigma$ we can represent this as $(2p_{zN} - f2p_{zO})$, where f is less than 1. The antibonding orbitals have a greater contribution from O 2p than from N 2p, so that the antibonding $2p\sigma^*$ can be represented as $(f2p_{zN} + 2p_{zO})$. Figure 4.28 shows orbitals of NO which are not symmetrical.

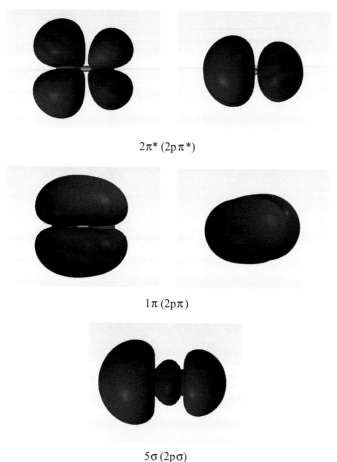

$2\pi^*$ $(2p\pi^*)$

1π $(2p\pi)$

5σ $(2p\sigma)$

Figure 4.28
Molecular orbitals of NO.

On Figure 4.27 we have labelled the molecular orbitals in the two ways that we introduced for homonuclear diatomic molecules. Note that we reach higher numbers when labelling the σ and π orbitals (for example $2\pi^*$ and 5σ) than we did for N_2 or O_2. This is because we can no longer distinguish g and u, and so have to label all σ (or π) orbitals consecutively.

From Figure 4.27, what is the bond order of NO?

It is $2\frac{1}{2}$. The 1s and 2s orbitals form bonding and antibonding pairs, each with two electrons, and so do not contribute to the bond order. There are six electrons, i.e. three pairs, in the bonding orbitals formed by the 2p orbitals, and one electron, or half a pair, in the antibonding orbitals $2\pi^*$, giving a bond order of $3 - \frac{1}{2} = 2\frac{1}{2}$.

How would you expect the bond length and dissociation energy of NO to compare with those of N_2 and O_2?

N_2, NO and O_2 all use similar orbitals for bonding. Their bond orders are 3, $2\frac{1}{2}$ and 2, respectively, and so we would expect the bond length and dissociation energy of NO to lie between those of N_2 and O_2. The experimental data for the three molecules given in Table 4.1 support this prediction.

Table 4.1 A comparison of O_2, NO and N_2

Property	O_2	NO	N_2
bond order	2	2.5	3
number of unpaired electrons	2	1	0
dissociation energy/kJ mol^{-1}	498	632	945
bond length/pm	121	115	109.4

The odd electron in NO is shown in Figure 4.27 as occupying an antibonding orbital, $2\pi^*$. This suggests that removal of this electron might be fairly easy.

What bond order would NO^+ have?

Three, the same as N_2. We would expect the NO^+ ion to have a very large dissociation energy like N_2, larger than that of NO.

Is there any evidence for NO^+? Yes. The nitrosyl halides will react with other halides to form crystalline salts of NO^+; for example

$$NOCl(g) + SbCl_5(l) = NO^+ SbCl_6^-(s)$$
$$NOF(g) + BF_3(g) = NO^+ BF_4^-(s)$$

These salts can be isolated and their structures determined by X-ray diffraction. They are strong oxidizing agents and react with water, so they cannot be studied in aqueous solution, although NO^+ is present in mixtures of concentrated sulfuric and nitric acid. Under certain circumstances, NO will also bind to metal ions, and in many of these compounds will bond as NO^+.

Molecular orbital theory then explains why NO is stable with respect to N and O atoms, and predicts that NO^+ will form fairly easily. We have not explained why nitric oxide gas consists of NO molecules rather than N_2O_2 molecules, or why NO is unstable with respect to oxidation by oxygen to NO_2. To do this, we would have to consider orbital energy-level diagrams for N_2O_2 or NO_2 and O_2 as well.

The other molecule that we shall study in this Section is CO. This molecule is formed when fossil fuels are incompletely burnt and is responsible for deaths from faulty gas heaters and car exhaust fumes. It can bind to haemoglobin in the blood thus preventing oxygen being transported to where it is needed in the body.

As we saw for C_2, the carbon 2s and 2p orbitals are fairly close in energy. For CO, then, we would expect that when we combine the 2p atomic orbitals on C and O to form molecular orbitals, we shall have to consider the effect of contributions from the carbon 2s. We do not need to consider O 2s as this is not close enough in energy to C 2s, C 2p and O 2p.

When we combine the 2p atomic orbitals from C and O, we obtain six molecular orbitals (5σ, 1π (two), 2π (two) and 6σ), as we did by combining 2p orbitals in NO. What effect will mixing in some carbon 2s orbital have?

In C_2, the effect of the 2s orbital on the molecular orbitals made from the 2p atomic orbitals was to raise the energy of the $3\sigma_g$ ($2p\sigma_g$) orbital so that it was above the $1\pi_u$ ($2p\pi_u$) orbital. Similarly, the effect of the 2s orbital in CO is to raise the energy of the 5σ (2pσ) orbital above that of the 1π (2pπ).

In other words, the 5σ orbital is no longer simply a combination of 2p on carbon and 2p on oxygen, but contains some carbon 2s orbital as well.

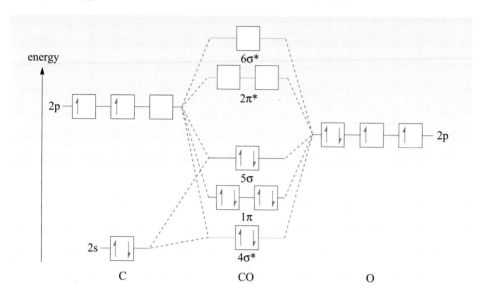

Figure 4.29
Orbital energy-level diagram for CO.

Molecular orbital theory predicts that CO has a very similar electronic structure to N_2, with which it is **isoelectronic**; that is, CO and N_2 have the same number of electrons (14). CO and N_2 have different bond lengths and molar bond enthalpies (113 pm and 110 pm, and 1 076 kJ mol⁻¹ and 945 kJ mol⁻¹, respectively) because the orbitals are not exactly the same in the two molecules. The values are, however, much closer to each other than they are to bond lengths and enthalpies of molecules with different bond orders.

The electrons in the molecular orbitals of CO, unlike those in N_2, are not evenly distributed between the atoms. Four electrons are transferred from O 2p orbitals to molecular orbitals that are mixtures of O 2p and C 2p and C 2s orbitals. Two electrons are transferred from C 2p orbitals to these mixed molecular orbitals.

Figure 4.30 compares similar molecular orbitals in CO and N_2.

If the molecular orbitals of CO were equal mixtures of C 2p and O 2p, then the net result would be a transfer of electron density corresponding to one electron from oxygen to carbon because four electrons came from oxygen and two from carbon, but in the CO molecule all six would be equally shared between oxygen and carbon giving ⁻C≡O⁺. Because the O 2p orbitals are lower in energy than the C 2p, the bonding molecular orbitals contain more O 2p than C 2p; this reduces the amount of electron transfer.

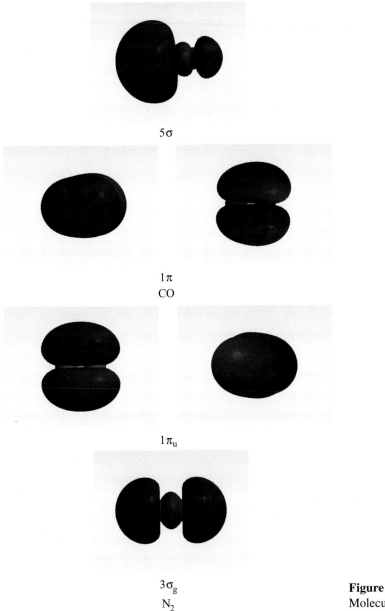

5σ

1π

CO

$1\pi_u$

$3\sigma_g$

N_2

Figure 4.30
Molecular orbitals for N_2 and CO.

We find experimentally that there is almost no transfer of electron density from C to O or from O to C. The dipole moment* is very small and has the positive end on the oxygen.

In its complex with haemoglobin, CO is bound to a metal atom (iron) using the 5σ and $2\pi*$ orbitals. Dinitrogen, however, despite its similar molecular orbital description, does not bind to iron in haemoglobin. The $1\pi_g$ orbital in N_2 is of a less suitable energy for binding to the iron atom than the $2\pi*$ orbital in CO. Dinitrogen does, however, bind to metal ions in complexes, and attachment of N_2 to a metal ion occurs in enzymes that are involved in making atmospheric nitrogen available to plants.

---

* Dipole moments are discussed in the CD-ROM accompanying this Book.

4.5 Summary of Section 4.4

1   Molecular orbitals for heteronuclear diatomic molecules can be made by combining atomic orbitals of the same symmetry and of similar energy from the two atoms.

2   Combination of two atomic orbitals of the same types (e.g. s, p_z, p_x) produces a bonding orbital of lower energy than either atomic orbital, and an antibonding orbital of higher energy. Further atomic orbitals can be mixed in by combining them with the resulting orbitals, provided they are close in energy and form the same type of molecular orbital (e.g. σ).

3   Because the atomic orbitals combined in heteronuclear diatomic molecules are not of the same energy, the electron in the molecular orbital is not equally shared between the two atoms.

4   Molecular orbitals in heteronuclear diatomic molecules cannot be labelled with the subscripts g and u because the molecule does not possess a centre of symmetry.

QUESTION 4.8

Draw an orbital energy-level diagram for OF, showing only those orbitals made from the 2p atomic orbitals on both atoms. Sketch the occupied molecular orbital that is highest in energy. What is the bond order of OF? Why is this molecule not observed under normal conditions? (F 2p orbitals are lower in energy than O 2p orbitals.)

QUESTION 4.9

Draw an orbital energy-level diagram for CN. Start by combining 2p orbitals on C and N, and then consider whether you need to include other orbitals. (The energy of the N 2p orbital lies between those of C 2p and C 2s. That of the N 2s is much lower than C 2s.) CN is known only from spectroscopic studies, but the cyanide ion CN$^-$ is very well known. With which molecule discussed in this Section is it isoelectronic? Does molecular orbital theory suggest that CN$^-$ might bind to iron?

POLYATOMIC MOLECULES

5

The same principles that we used to build up molecular orbitals for diatomic molecules can be applied to larger and, indeed, very large molecules. We shall consider a few examples and then some general points that apply to calculations on any molecule. For our first example, we pick up on the molecules that we only partially dealt with using hybrid orbitals.

5.1 Unsaturated organic molecules

When we formed hybrid orbitals for ethene, $H_2C=CH_2$, we had one 2p orbital on each carbon left over. These p orbitals have lobes above and below the plane of the molecule as in Figure 5.1.

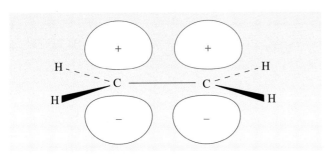

Figure 5.1
p orbitals in ethene.

● Look back at the orbitals for diatomic molecules and suggest what type of molecular orbital these 2p orbitals might form.

● They could overlap to form a bonding orbital similar to the π orbitals in diatomic molecules.

● How many bonding orbitals of this type could be formed in ethene?

● Only one. There is only one pair of 2p orbitals left over for this type of bond. This bonding orbital is shown in Figure 5.2.

In an extension to the labelling used for diatomic molecules, this orbital is referred to as a π orbital. We can describe the bonding in ethene as consisting of bonding orbitals in the plane of the molecule formed by overlap of the sp^2 hybrid orbitals with each other and with the s orbitals on hydrogen and an out-of-plane π-bonding orbital. By analogy with orbitals for diatomic molecules, the orbitals formed by the sp^2 hybrid orbitals are called σ orbitals.

Figure 5.2
π orbital in ethene.

As in diatomic molecules, π bonding is weaker than σ bonding. The reactivity of ethene relative to ethane can thus be traced to the relative ease with which the π bond can be broken.

Simple π bonds of this type are also found in molecules with carbonyl groups — aldehydes, ketones, carboxylic acids and esters (Structures **5.1**).

$$R^1 \diagdown C=O \diagup H$$
aldehyde

$$R^1 \diagdown C=O \diagup R^2$$
ketone

$$R^1 \diagdown C=O \diagup HO$$
carboxylic acid

$$R^1 \diagdown C=O \diagup R^2O$$
ester

5.1

In these cases the π orbital is formed by overlap of a 2p orbital on carbon with one on oxygen.

In ethyne, there are two π bonding orbitals because the formation of sp hybrids leaves two p orbitals on each carbon. The bonding in ethyne is very similar to that in dinitrogen; it has two filled π bonding orbitals as in dinitrogen and a filled σ bonding orbital (Figure 5.3). The σ bonding orbital, however, is not confined to the

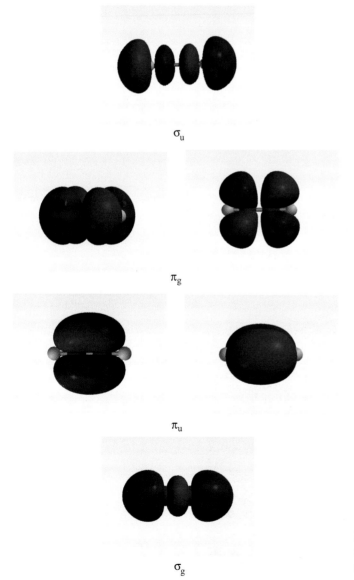

σ_u

π_g

π_u

σ_g

Figure 5.3
σ_g and π_u bonding, and π_g and σ_u antibonding orbitals in ethyne, C_2H_2.

carbon atoms but includes the hydrogen as well. This orbital together with two other σ orbitals can be thought of as producing the carbon–hydrogen and carbon–carbon bonds formed from sp hybrids on carbon and 1s on hydrogen.

Where a molecule has two or more double bonds separated by single bonds (a **conjugated molecule**), the p orbitals on several atoms can overlap to form **delocalized** π bonding **orbitals**. In buta-1,3-diene, $H_2C=CH-CH=CH_2$, p orbitals on all four carbons can overlap to form a π orbital covering the entire carbon framework (Figure 5.4).

Buta-1,3-diene also has an occupied π bonding orbital (Figure 5.5) in which p orbitals on carbons 1 and 2 overlap, as do those on carbons 3 and 4, but those on carbon 2 do not overlap in a bonding fashion with those on carbon 3. π bonding is thus stronger between carbons 1 and 2 and between carbons 3 and 4 than between carbons 2 and 3, and this is reflected in the bond lengths (Figure 5.6).

The most famous example of such a delocalized orbital is the one in benzene, C_6H_6, in which 2p orbitals on all six carbons overlap forming rings of electron density above and below the benzene plane (Figure 5.7 overleaf). There are also other occupied π orbitals in benzene in which two pairs or two triplets of atoms are

Figure 5.4 π orbital in buta-1,3-diene. 🖳

Figure 5.5 Occupied π orbital in buta-1,3-diene.

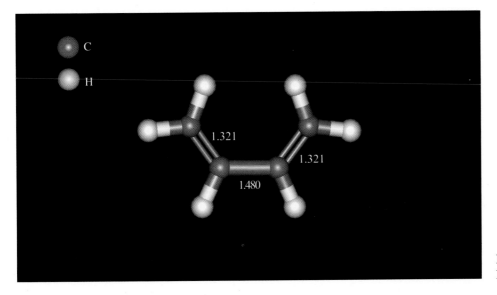

Figure 5.6
Bond distances in buta-1,3-diene. 🖳

(a)   (b)

Figure 5.7 π orbitals in benzene. (a) orbital covering all six atoms 🖥, (b) other occupied π orbitals.

bonded. Together these also give equal electron occupation on each of the six carbons. Thus the electron density is identical for all six carbons of the benzene ring and all the carbon–carbon bond lengths are the same. Benzene can be written as resonance structures which reflect this delocalization (Structure **5.2**) and which indicate that all bonds are equivalent. Butadiene is not normally drawn as resonance structures; if we wish to do so to indicate the delocalization of the π orbitals over the C2—C3 bond, we end up with odd electrons on the outer carbons which we can leave as such (Structure **5.3**) or pair up on one carbon (Structure **5.4**).

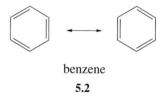

benzene

5.2

butadiene

5.3

or

5.4

The left-hand part of both Structures **5.3** and **5.4** will contribute more to the resonance hybrid than the others. Benzene is thus a delocalized hydrocarbon whereas buta-1,3-diene with its alternating carbon–carbon bond lengths is a conjugated hydrocarbon.

Propadiene, $H_2C=C=CH_2$, unlike buta-1,3-diene does not form a delocalized π orbital. Why does it not do so?

The central carbon is sp hybridized and has two free p orbitals. One of these overlaps with a p orbital on one terminal carbon and the other with a p orbital on the other terminal carbon. However, these two π bonds are at right-angles and so do not interact with other.

⬤ Will penta-1,4-diene form a delocalized π orbital?

⬤ No. The two double bonds in this molecule are separated by an sp^3 hybridized carbon with no free p orbitals: $H_2C=CH-CH_2-CH=CH_2$.

QUESTION 5.1

Sketch a delocalized π bonding orbital for cyclohexa-1,3-diene.

QUESTION 5.2

Which of the following molecules will have a delocalized π orbital:

(i) but-1-ene   (ii) hexa-2,4-diene   (iii) cyclohexa-1,4-diene?

5.2 Delocalized π orbitals in inorganic molecules

Unsaturated molecules are less common in inorganic chemistry but π bonding orbitals of the type we have just introduced are found. One simple example is ozone, O_3. Ozone is a pollutant at ground level but further out in the atmosphere protects us from the Sun's ultraviolet radiation.

O_3 is a V-shaped molecule. Two of the 2p orbitals on each oxygen form σ-bonding orbitals and the remaining one, which has lobes above and below the plane of the molecule, forms a delocalized π bonding orbital (Figure 5.8). It is possible to form such an orbital because the central oxygen atom uses only one p orbital to π-bond, whereas in propadiene the central carbon used two p orbitals. In ozone, the central oxygen can be regarded as sp^2 hybridized, forming two σ bonds and with a non-bonded pair of electrons occupying the third sp^2 orbital. Ozone adopts this bonding structure because oxygen has two more valence electrons than carbon and these could not be accommodated in a propadiene-like structure. Ozone has a σ bond order of two and a π bonding orbital that covers all three atoms and contains two electrons, giving a total bond order of three. Thus we can think of each bond as having an order of $1\frac{1}{2}$. This ties in with the simple theory of bonding in which we write ozone as two resonance forms $O=O\rightarrow O \longleftrightarrow O\leftarrow O=O$, and with the observed O—O bond distance 128 pm, which is between that in the doubly-bonded O_2 (121 pm) and the O—O single bond in hydrogen peroxide, H_2O_2 (149 pm).

Figure 5.8 π orbital for ozone.

Another example is boron trifluoride. Boron trifluoride exists as BF_3 molecules so that the boron atom only has six valence electrons around it rather than a full octet. Boron has three valence electrons and forms three single bonds to fluorine. The molecule is stabilized by non-bonded pairs of electrons from fluorine occupying a delocalized π orbital formed from 2p orbitals on the three fluorines and on boron (Figure 5.9).

We can show this as the resonance hybrid, **5.5**:

Figure 5.9 π orbital in boron trifluoride.

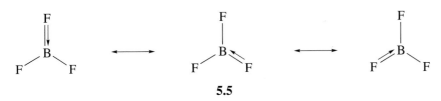

5.5

By contrast the hydride BH_3 is not found as BH_3 molecules but as B_2H_6 molecules.

● Why is BH_3 not stabilized by π bonding?

● Hydrogen has only one electron and so after forming the single bond to boron would have no non-bonding pairs to donate to boron.

5.3 Electron-deficient molecules and three-centre bonds

Electron-deficient molecules are those that have insufficient valence electrons to form the conventional two-centre, two-electron bonds required by the molecular structure. An example is diborane, B_2H_6, Structure **5.6**.

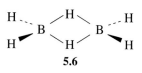

5.6

This has a structure in which the two borons each form two boron–hydrogen single bonds and are linked by two hydrogen bridges. The bridging hydrogens have only one valence electron and one valence orbital, 1s, and yet they are bonded to two borons. How?

You have seen that a pair of electrons in a delocalized π orbital can bond three atoms together (as in O_3). In diborane a single σ orbital binds the two boron atoms to a bridging hydrogen. This orbital can be thought of as being formed by the overlap of the 1s on hydrogen with sp^3 hybrid orbitals on both borons. In Figure 5.10 you can see the electron density around the hydrogen atoms showing them equally bonded to each boron. Boron forms a large number of hydrides in which not only BHB but also BBB three-centre bonds occur.

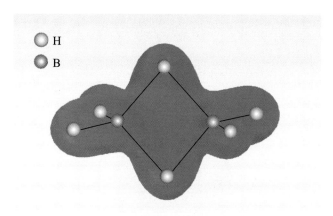

Figure 5.10
Electron density in diborane, showing the BHB bridges.

Three-centre bonds are also found in compounds, particularly of sulfur, the halogens and the noble gases in which there is formally an expansion of the octet (that is, the central atom is surrounded by more than eight electrons in the Lewis structure). For example, the bonding in XeF_2 can, to a first approximation, be attributed to a σ-bonding orbital formed from p orbitals on all three atoms as in Figure 5.11. Such an orbital resembles two Xe—F bonds but is only occupied by two electrons.

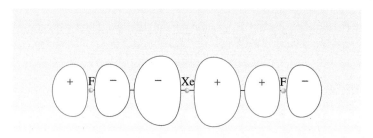

Figure 5.11
Three-centre σ orbital in $XeFe_2$.

5.4 Bonding in the water molecule

In the previous two Sections we only considered some of the molecular orbitals in the molecules we discussed. We now look briefly at the complete set of molecular orbitals for a simple polyatomic molecule, H_2O.

Valence-shell electron-pair repulsion (VSEPR) theory and the concept of hybridization suggest that the water molecule has two O—H bonds and two non-bonded pairs arranged tetrahedrally. More accurate calculations show that this does not provide a true picture of the total electron density in H_2O.

The lowest-energy orbital in H_2O is the 1s on oxygen. This is so far removed in energy and overlaps so little with the other atomic orbitals that it can be regarded as a **non-bonding orbital**, that is one that is neither bonding nor antibonding. Non-bonding orbitals like the oxygen 1s often retain their atomic orbital character. The 2s orbital on oxygen plays a minor role in bonding. To a first approximation we can also regard this as non-bonding. The two 2p orbitals on oxygen with their lobes in the plane of the molecule form σ-bonding orbitals with the 1s orbitals on hydrogen. These are shown in Figure 5.12. The third 2p orbital on oxygen would be available to form a π bond (Figure 5.13), but the only orbitals available on hydrogen are 1s orbitals and these will not bond in this manner. The electrons in this 2p orbital do not contribute to the bonding; the orbital is non-bonding.

(a)

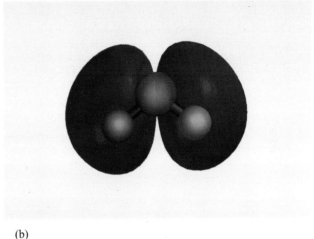

(b)

Figure 5.12 σ-bonding orbitals in the water molecule. 🖳

Figure 5.13
π orbital in water.

The two bonding orbitals together do give electron density where you would expect it for two O—H bonds but also on the far side of the oxygen atom from the hydrogens, where it is enhanced by the 2s orbital on oxygen. Figure 5.14 shows the total electron density around H_2O, and you can see the considerable density beyond the oxygen. This is part of what we would regard as the non-bonding pairs on oxygen. The remaining non-bonding pair electron density can be seen in the perpendicular view shown in Figure 5.15. Note that this non-bonding electron density does not form two lobes resembling potential bonds as you might have expected from the hybrid orbital description.

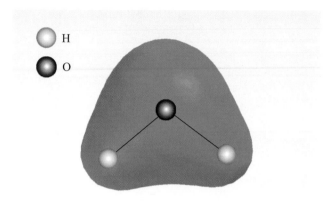

Figure 5.14 Electron density in the H_2O molecule.

Figure 5.15 Electron density in the H_2O molecule shown with the plane of the molecule at right-angles to the page.

The orbitals we have discussed can be shown on a molecular orbital energy-level diagram as was done for diatomic molecules. If we only consider the 2p orbitals on oxygen and the 1s on hydrogen, then the lowest-energy orbital is one of those in Figure 5.12. Which of these two is lower in energy? Both orbitals are combinations of the same types of atomic orbital; the difference in energy will depend on the difference in overlap. If water were a linear molecule, then one 2p orbital would overlap well with the H 1s orbitals but the other would not overlap at all. As the molecule bends, the ∠HOH angle is reduced and the overlap of the first orbital decreases but the other 2p orbital starts to overlap. For H_2O in its equilibrium geometry with an angle of 104.5°, the orbital shown in Figure 5.12b is the lower in energy.

Next in energy above the two bonding orbitals lies the non-bonding 2p orbital on oxygen (Figure 5.13). Finally there are antibonding orbitals corresponding to the bonding orbitals. The complete diagram is shown in Figure 5.16.

QUESTION 5.3

Use the method introduced for diatomic molecules to determine the total bond order of H_2O. Non-bonding orbitals do not contribute.

QUESTION 5.4

Figure 5.17 shows a molecular orbital energy-level diagram for ammonia. Label the molecular orbitals as bonding, non-bonding or antibonding. What is the total bond order?

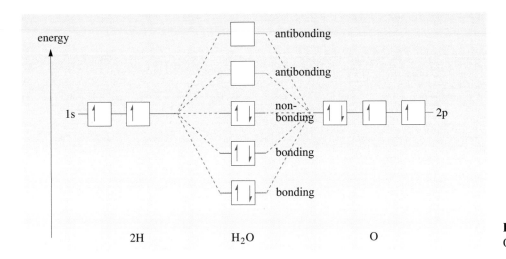

Figure 5.16
Orbital energy-level diagram for H_2O.

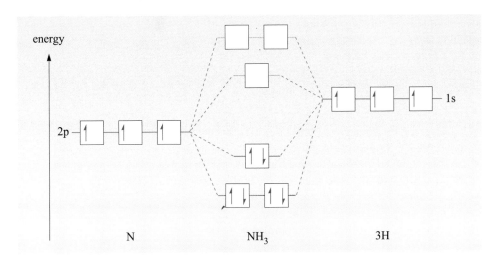

Figure 5.17
Molecular orbital energy-level diagram for ammonia, NH_3.

5.5 Summary of Section 5

1   Molecular orbitals for polyatomic molecules can be constructed using the same principles (overlap of similar energy atomic orbitals) as for diatomic molecules.

2   In unsaturated organic molecules, the p orbitals on carbon not used in hybridization can form π bonds.

3   Where there is more than one double bond in an organic molecule and two or more of the double bonds are separated by just one single bond — that is, the molecule is conjugated — a delocalized π orbital is formed.

4   Delocalized orbitals, covering more than two atoms, also occur in some inorganic molecules.

5   Molecular orbitals for H_2O can be formed from 2p orbitals on oxygen and 1s orbitals on hydrogen.

6   The energy of the bonding molecular orbitals in H_2O depends on the $\angle HOH$ angle.

7   Molecular orbital energy-level diagrams similar to those for diatomic molecules can be constructed for polyatomic molecules.

SYMMETRY AND MOLECULAR ORBITAL CALCULATIONS

6

Calculations to obtain orbitals and their energies for small and medium-sized molecules are routinely performed by a number of computer packages. As described in Section 4, these build up molecular orbitals by finding the combinations of atomic orbitals that give the lowest energy. With diatomic molecules, you saw that we could label the molecular orbitals σ or π. However, in polyatomic molecules, although it is useful to refer to σ and π bonds, the labels we give to the molecular orbitals and their energy levels depend on the symmetry of the molecule.

The use of symmetry reduces the number of calculations required. In water, for example, we know that if the 1s orbital on one hydrogen is involved in an orbital then the 1s orbital on the other hydrogen is equally involved (although it may have a different sign). We do not therefore need to consider the two hydrogens independently. In a highly symmetrical molecule such as benzene, where all six carbons are equivalent, this leads to a considerable saving in computer time. The symmetry of the molecule also determines the degeneracy of the molecular orbitals — that is, how many must have the same energy — and this aspect is important in transition metal chemistry.

Chemists have adopted from mathematicians a way of classifying molecules according to their symmetry. The classification is based on the idea that if a molecule is sufficiently symmetrical, then an action can be found that will leave the molecule looking the same as it did when you started. You have already met two such actions: inversion through a centre of symmetry and rotation about the molecular axis of a diatomic molecule. Actions such as this are called **symmetry operations**, and a molecule can be placed in a category according to how many such operations you can perform and still leave the molecule looking the same.

It is sometimes difficult to spot symmetry operations and we recommend that you use a model kit in this Section to visualize the molecules in 3-D.

6.1 Rotation

The objects in Figure 6.1 have something in common. If asked what it was, you might suggest it has something to do with three — the objects have three points or three spokes or three sides etc. This is correct, but in order to use the symmetry classification, we have to find an action that will leave all the objects looking the same. Now, if they are going to look the same, each spoke, point, line, etc., must move to the position of the next one. One thing we could do is turn each object round so that each corner of the triangle, for example, has moved round to the position of the next one. If the corners of the triangle in Figure 6.1 are labelled a, b and c then the result of this action is shown in Figure 6.2. The same action will produce an object identical with the one you started with when performed on all the objects in Figure 6.1. So the action is a symmetry operation. What we have done is rotate the object through one-third of a complete revolution. Such an action is a rotation, and because three such rotations return the object to its starting position, it is known as a threefold rotation.

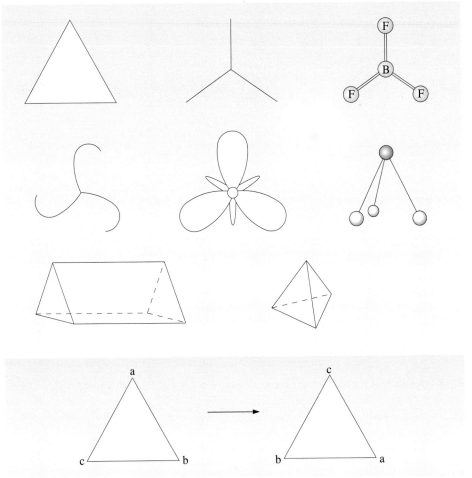

Figure 6.1
A set of objects.

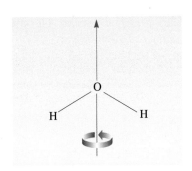

Figure 6.2
The result of rotating an equilateral triangle through a third of a revolution.

For a complete description of the symmetry operation, we need to specify what the objects are rotating about. If you consider a wheel on a cart or a buggy, the wheel rotates about an axle. We can imagine a sort of axle going through the objects in Figure 6.1. This is illustrated for the BF_3 molecule in Figure 6.3. The line about which the molecule is rotated is called an **axis of symmetry**, in this case a threefold axis of symmetry. An axis of symmetry is an example of a **symmetry element**.

The objects in Figure 6.1 had a threefold axis of symmetry; other objects need to be turned through different amounts to remain identical. Water, H_2O, for example, has a twofold axis of symmetry (Figure 6.4). Rotation through half a complete revolution about the line shown produces an identical-looking molecule.

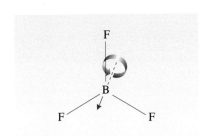

Figure 6.3
A threefold axis in BF_3.

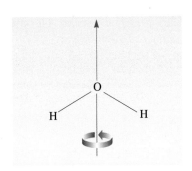

Figure 6.4
The twofold axis of symmetry in H_2O.

Both square-planar molecules such as XeF_4 and octahedral molecules such as SF_6 have fourfold axes. A quarter turn produces an identical-looking molecule, and four such turns are needed to take the molecule back to its starting point. The fourfold axis of XeF_4 and one of the fourfold axes of SF_6 are shown in Figure 6.5.

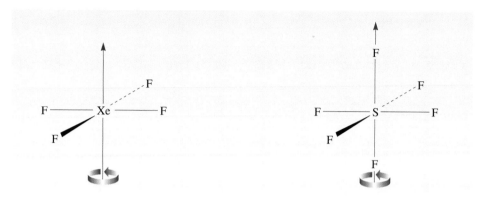

Figure 6.5
Fourfold axes in XeF_4 and SF_6.

🔘 Has ammonia, NH_3, an axis of symmetry?

🔘 Yes: it has a threefold axis of symmetry, which goes through the nitrogen atom. This is shown in Figure 6.6. You may find it helpful to build a model of NH_3 and rotate it through a third of a revolution.

Both BF_3 and NH_3 have threefold axes but the molecules are not the same shape. BF_3 is trigonal planar and NH_3 is pyramidal. Our classification scheme allows us to distinguish these two shapes by considering what other symmetry operations might apply to the two. The first thing to note is that molecules (or other objects) may have more than one axis of rotation.

Suppose we rotate BF_3 about a B—F bond. After half a turn we arrive at something that looks identical with the starting molecule. Two such turns are required to restore us to the starting point and so the B—F bond is a twofold axis (Figure 6.7).

There are three B—F bonds and no reason for choosing one rather than another, so each B—F bond is a twofold axis. BF_3 therefore has three twofold axes, all of which are at right-angles to the threefold axis.

A situation like this in which a molecule has an *n*-fold axis of symmetry and *n* twofold axes at right-angles is not uncommon, and so it is worth looking for *n* twofold axes when you have found an *n*-fold axis (*n* is 2 or higher).

Rotating NH_3 about an N—H bond, by contrast, produces the result shown in Figure 6.8.

The N—H bonds in NH_3 are not axes of symmetry. NH_3 in fact has only one axis of symmetry, the threefold axis already discussed. The shapes of NH_3 and BF_3 can thus be distinguished by the number of symmetry axes.

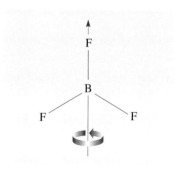

Figure 6.6
The threefold axis in NH_3.

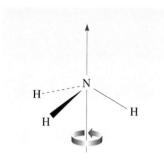

Figure 6.7
A twofold axis in BF_3.

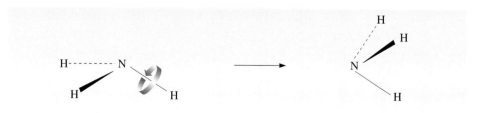

Figure 6.8
Rotation of NH_3 by half a revolution about the N—H bond indicated.

● How many symmetry axes has XeF_4?

● Earlier we said that XeF_4 has a fourfold axis. We look to see if there are any others. A good starting place is to see if there are n twofold axes at right-angles to the n-fold axis. In this case $n = 4$ and so we look for four twofold axes. The fourfold axis is at right-angles to the plane of the molecules (Figure 6.5) and so these twofold axes, if they exist, must be in the plane. An axis running through two opposite Xe—F bonds is a twofold axis (Figure 6.9a), and there are two such axes. There are another two axes in this plane, lying between Xe—F bonds (Figure 6.9b). So XeF_4 has one fourfold axis and four twofold axes.

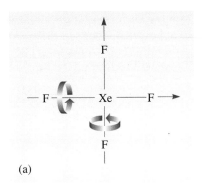

(a)

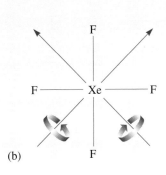

(b)

Figure 6.9
(a) Twofold axes along Xe—F bonds in XeF_4. (b) Twofold axes between Xe—F bonds in XeF_4.

There cannot be an axis of higher order than fourfold in XeF_4 because there are only four fluorine atoms and, at each turn, an atom must move to a position occupied by an atom identical with itself. Is there a threefold axis? Well, a threefold axis would interchange three fluorines and so such an axis would have to lie along an Xe—F bond. But the Xe—F bonds are only twofold axes and so there is no threefold axis.

One fourfold axis and four twofold axes are thus the total for XeF_4.

> As a general strategy for molecules with a central atom surrounded by n atoms of the same type, you first of all divide the n atoms into sets occupying similar positions. If the largest number of atoms in a set is m, then the highest-order axis you need to look for is m-fold. You may not find an m-fold axis; there are some molecules (for example, CH_4 and SF_6) for which, although all n atoms are in one set, there is no n-fold axis. This is because the atoms cannot be interchanged by rotation unless they all lie in a plane. The three hydrogens in ammonia lie in a plane but the four hydrogen atoms in methane do not. If you cannot find an m-fold axis, look for a lower order one ($m - 1$, $m - 2$, etc.).
>
> When you have found an axis of order 2 or higher, look for twofold axes at right-angles to this axis.

There is one special case that we must consider, and that is linear molecules. If we take CO_2 there are only two oxygens, so normally we would look only for a twofold axis. But in the case of linear molecules, there is a symmetry axis running along the molecular axis. Now, if we rotate the molecule about this axis, it remains looking the same however much we turn it. Because we could rotate it by an infinitesimal amount, we call this axis an infinite axis of symmetry. All linear molecules have such an axis. As with non-linear molecules we still have to look for n twofold axes at right-angles to the infinite axis, and in this case it means looking for an infinite number of such axes. In CO_2 there are an infinite number of twofold axes, and Figure 6.10a illustrates one of them. We have drawn this axis in the plane of the paper, but any axis at right-angles to the molecular axis at any angle to the plane of the paper will be a twofold axis as well. Figure 6.10b shows the molecule viewed along its infinite axis, and three of the infinite number of twofold axes.

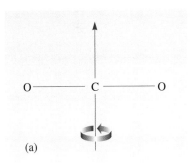

(a)

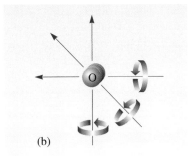

(b)

Figure 6.10
Twofold axes in CO_2.

● What are the axes of symmetry in a less symmetrical linear molecule, such as HCN?

● HCN will have an infinite axis because it is linear. It does not, however, have any twofold axes because there are no identical atoms to interchange.

Now try the following questions to practise finding axes of symmetry, and then go on to see what other symmetry operations there are.

QUESTION 6.1

Find all the axes of symmetry in the following molecules, and state whether they are twofold, threefold, etc.: (a) BrF_5; (b) SO_2; (c) $POCl_3$; (d) NO_3^-; (e) HBF_2. (The shapes of these molecules can be obtained using VSEPR theory.)

QUESTION 6.2

Are the axes (a)–(d) axes of symmetry? If so, what is their order (twofold, threefold...)?

(a) A line along one C—F bond in CF_4.

(b) A line through the centre of a benzene ring, Structure **6.1**, perpendicular to the plane of the molecule.

(c) A line in the plane of an ethene molecule at right-angles to the C=C bond and going through the centre of this bond (Figure 6.11a).

6.1

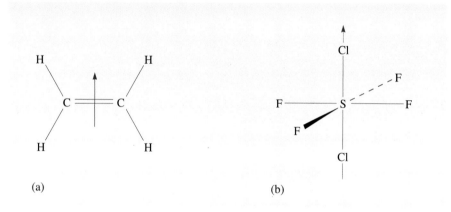

(a)

(b)

Figure 6.11
(a) An axis in ethene. (b) An axis in SCl_2F_4.

(d) A line through the two S—Cl bonds in SCl_2F_4, with the two chlorines opposite (Figure 6.11b).

(e) A line through the three carbons in propadiene, Structure **6.2**:

6.2

6.2 Reflection

The molecule NFH_2, Structure **6.3**, looks more symmetrical than NFHCl, Structure **6.4**, but neither molecule has an axis of symmetry. Is there a symmetry operation that will distinguish these two?

6.3   **6.4**

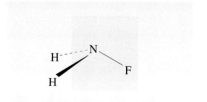

Figure 6.12
A symmetry plane in NFH_2.

Yes, it is called reflection. If we draw a plane that contains the N—F bond of NFH_2 and lies between the two N—H bonds as in Figure 6.12, then a line drawn from one hydrogen atom to the plane and out the other side by the same amount meets the other hydrogen atom. The reflection of each hydrogen in the plane is the other hydrogen. This is our next symmetry operation — reflection through a **plane of symmetry**.

NFHCl has no plane of symmetry, and so this distinguishes it from NFH$_2$.

NFH$_2$ has only a plane of symmetry, but many molecules have both plane(s) and an axis (or axes) of symmetry. NH$_3$, for example, has a plane of symmetry just like that in Figure 6.12. Because all three atoms bonded to nitrogen are the same in NH$_3$, however, it does not matter which N—H bond lies in the plane and so there are three such planes (Figure 6.13).

NH$_3$ has a threefold axis of symmetry as well as the three planes of symmetry. The three planes each contain this axis. NH$_3$ has only one axis of symmetry, and so this threefold axis is the **principal axis** of NH$_3$, the axis of highest order. Planes of symmetry containing the principal axis are known as **vertical planes**. As it is usual to denote the principal axis of the molecule as the z-axis and to have the z-axis pointing up the page, then such planes of symmetry would indeed be vertical when the molecule is oriented in the standard manner. NH$_3$, then, has one threefold axis of symmetry and three vertical planes of symmetry. Molecules with an n-fold principal axis and n vertical planes of symmetry are common, and so it is worth looking for these planes when you have found the principal axis.

Now consider BF$_3$. This molecule, like NH$_3$, has as its principal axis a threefold axis. Does it have three vertical planes of symmetry? Yes: as in NH$_3$, each fluorine lies in a plane of symmetry that reflects the other two fluorines into each other. We saw that BF$_3$ had more axes of symmetry that NH$_3$; has it more planes of symmetry than NH$_3$?

BF$_3$ is a planar molecule. What happens if we reflect each atom through this plane? The molecule is unchanged, because all the atoms lie in the plane and so do not move at all when reflected. The plane of the molecule is thus a plane of symmetry. However, this is not a vertical plane of symmetry. The principal axis of BF$_3$ is at right-angles to the plane; if we arrange the molecule with the principal axis vertical then the molecular plane is horizontal. A plane of symmetry at right-angles to the principal axis is known as a **horizontal plane**; BF$_3$ thus has three vertical planes of symmetry and one horizontal plane of symmetry.

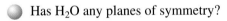 Has H$_2$O any planes of symmetry?

Yes. H$_2$O is planar and thus the molecular plane is a plane of symmetry. However, the principal axis of H$_2$O (the twofold axis — its only axis of symmetry) lies in the molecular plane. The molecular plane of H$_2$O is thus a vertical plane of symmetry. H$_2$O also has another plane of symmetry. This lies between the two O—H bonds and is also vertical. The two planes are shown in Figure 6.14.

The plane of symmetry in NFH$_2$ is neither vertical nor horizontal because NFH$_2$ has no axis of symmetry. It is referred to simply as a plane of symmetry.

It is useful to remember that *all planar molecules must have a plane of symmetry* — the plane of the molecule. Since all diatomic and triatomic molecules are planar, they all have at least one plane of symmetry. Planes of symmetry are of interest if you want to determine whether a molecule is chiral. A molecule cannot be chiral if it contains a plane of symmetry. For example, a substituted methane with two hydrogens substituted, such as CH$_2$FCl, has a plane of symmetry which runs

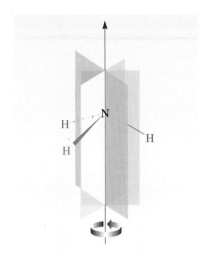

Figure 6.13
The three planes of symmetry in NH$_3$.

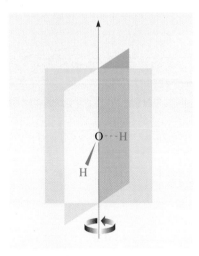

Figure 6.14
Planes of symmetry in H$_2$O.

through the carbon and the two substituents. In the case of CH_2FCl, the FCCl plane is a plane of symmetry. Planes of symmetry will also be present if there are three or four identical substituents. The only substituted methane without a plane of symmetry is one with four different groups around the carbon. This is the condition for such a molecule to be chiral.

The lack of a plane of symmetry as a criterion for potential optical activity allows us to identify other molecules that are chiral. For example, if one hydrogen on each terminal carbon of propadiene is substituted for then the resulting molecule is chiral. 1,3- Difluoropropadiene, for example, has a twofold axis of symmetry but no plane of symmetry. The enantiomers are shown in Figure 6.15. Construct models of these and convince yourself that they are non-superimposable.

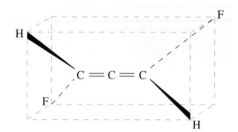 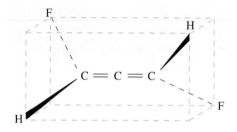

Figure 6.15
1,3-Difluoropropadiene.

QUESTION 6.3

The descriptions (i) to (vii) are of planes in molecules. For each one, decide whether it is (a) a vertical plane of symmetry, (b) a horizontal plane of symmetry, (c) a plane of symmetry that is neither vertical nor horizontal, or (d) not a plane of symmetry at all.

(i)   The molecular plane of HOF, Structure **6.5**:

(ii)   The molecular plane of *trans*-1,2-difluoroethene, Structure **6.6**:

(iii) A plane containing the P=O bond and one P—Cl bond in $POCl_3$.

(iv)   A plane through the carbon atom and at right-angles to the molecular axis in CO_2.

(v)   A plane containing the HF molecule.

(vi)   The plane of the three hydrogen atoms in NH_3.

(vii) The molecular plane of NO_3^-.

QUESTION 6.4

Optically active trialkylphosphines ($PR^1R^2R^3$) can be prepared. Using the criterion of a lack of a plane of symmetry, what can you say about the alkyl groups?

6.3 Inversion

The final symmetry operation in this Book is inversion through a centre of symmetry. You met this operation when you were studying homonuclear diatomic molecules, but it is not confined to diatomic molecules. Structure **6.7**, for example, has a centre of symmetry in the middle of the benzene ring:

A line from a bromine to the centre and out the other side brings us to a bromine, from a chlorine to a chlorine, a carbon to a carbon and a hydrogen to a hydrogen.

Square-planar molecules such as XeF_4 also have a centre of symmetry.

How about tetrahedral molecules? In tetrahedral molecules, such as CH_4, inversion through the carbon atom would turn the molecule inside out as in Figure 6.16. The molecule is not the same, and so this is not a symmetry operation. Tetrahedral molecules do *not* have a centre of symmetry. On the other hand, octahedral molecules do have a centre of symmetry.

Figure 6.16
Inversion of CH_4 through the C atom.

Lack of a centre of symmetry is also a property of chiral molecules.

● Which of the following molecules have a centre of symmetry: OF_2, F_2, HF, CO_2 and SF_6?

● F_2, CO_2 and SF_6 only.

You will recall that in homonuclear diatomic molecules, orbitals that were unchanged by inversion through the centre of symmetry were labelled g and those that were changed were labelled u. Orbitals of all molecules with a centre of symmetry can be labelled using g or u subscripts.

QUESTION 6.5

Do the following molecules have a centre of symmetry: (a) PF_5; (b) CO_3^{2-}; (c) ICl_2^- $(Cl-I-Cl)^-$; (d) ethene; (e) ethane in its staggered conformation?

QUESTION 6.6

The $CaCl_2$ molecules formed by heating solid calcium chloride has a centre of symmetry. Is it straight or V-shaped?

QUESTION 6.7

The complex anion $[PtCl_4]^{2-}$ has a centre of symmetry. What shape is it?

6.4 Symbols for symmetry elements

To save writing out the axis of symmetry, plane of symmetry or centre of symmetry in full each time we mention them, we give each of these **symmetry elements** a symbol. Thus a plane of symmetry is given the symbol σ (sigma) and a centre of symmetry the symbol i. An *n*-fold axis of symmetry is given the symbol C_n; that is, a twofold axis is signified by C_2, a threefold axis by C_3, and so on.

The corresponding symmetry operations (that is, the acts of rotation, reflection, etc.) are given the same symbols, but we shall distinguish them by putting a circumflex accent over the symbol for an operation, for example $\hat{\sigma}$. Table 6.1 summarizes the symmetry elements and operations that you have met so far. If a molecule possesses more than one plane of symmetry or n-fold axis of symmetry, we can indicate this by putting the appropriate number before the symbol for the symmetry element. Thus, if the molecule has four twofold axes (for example XeF_4), we write $4C_2$. Vertical and horizontal planes of symmetry are distinguished by subscripts, σ_v and σ_h.

Let's finish this Section by listing all the symmetry elements of H_2O.

● How many axes of symmetry has H_2O?

● One twofold axis.

● How many planes of symmetry has H_2O?

● We saw in Section 6.2 that it had two.

● Are these planes vertical or horizontal?

● Vertical.

● Does the molecule have a centre of symmetry?

● No.

The complete list using the symbols just introduced is then $C_2 + 2\sigma_v$.

Table 6.1 Symbols for symmetry elements and symmetry operations

Symmetry element	Symbol	Symmetry operation	Symbol
plane of symmetry	$\sigma, \sigma_v, \sigma_h$	reflection through the plane	$\hat{\sigma}, \hat{\sigma}_v, \hat{\sigma}_h$
n-fold axis of symmetry	C_n	rotation through $1/n$ of a complete revolution about the axis of symmetry	$\hat{C}_n$
centre of symmetry	i	inversion through the centre of symmetry	$\hat{i}$

6.5 Summary of Sections 6.1–6.4

1   The symmetry of a molecule or other object can be described by saying how many and which symmetry elements it contains.

2   Each symmetry element has associated with it a symmetry operation.

3   If an object contains a particular symmetry element, then the action of the associated symmetry operation will produce an identical-looking object occupying the same position in space as the original.

4   Important symmetry elements are a plane of symmetry, σ, an axis of symmetry, C_n, and a centre of symmetry, i.

5   The associated symmetry operations are reflection through the plane of symmetry, $\hat{\sigma}$, rotation by $1/n$ of a complete revolution about the axis of symmetry, $\hat{C}_n$, and inversion through the centre of symmetry, $\hat{i}$.

6   Molecules that possess a plane or centre of symmetry cannot be chiral.

QUESTION 6.8

List all the symmetry elements of the molecules (a) BF_3, (b) SF_4 and (c) ethene, C_2H_4, and give them their correct symbols. (Use VSEPR theory to predict the shapes.)

6.6 Symmetry point groups

We have looked at several molecules to see what symmetry elements they contain. Now we are going to meet a way of classifying molecules (and other objects) according to their symmetry. What we do is to group together all molecules and other objects that contain the same symmetry elements. The symmetry operations of any molecule form a mathematical group called a **symmetry point group**. All molecules with the same symmetry elements will be unchanged by the same symmetry operations, and hence are said to belong to the same symmetry point group, denoted by a particular symbol. The labelling of molecular orbitals is determined by the symmetry point group to which the molecule belongs.

H_2O and all objects that contain one twofold axis of symmetry and two vertical planes of symmetry belong to the same group, denoted by the symbol $\mathbf{C_{2v}}$ (pronounced see-two-vee).

Let us now look at another group, the group of molecules with the same symmetry elements as NH_3.

⬤ What symmetry elements does NH_3 contain?

⬤ One threefold axis and three vertical planes of symmetry.

The group of molecules with the same symmetry elements as H_2O was labelled $\mathbf{C_{2v}}$ because the axis of highest symmetry was twofold, and the two planes of symmetry were vertical.

⬤ Can you suggest a symbol for the group of molecules with the same symmetry elements as NH_3?

⬤ The symmetry point group is labelled $\mathbf{C_{3v}}$.

The members of the group $\mathbf{C_{2v}}$ have symmetry elements $C_2 + 2\sigma_v$. Those in $\mathbf{C_{3v}}$ contain the elements $C_3 + 3\sigma_v$. What symmetry elements would expect molecules in a group labelled $\mathbf{C_{4v}}$ to contain? By comparison with $\mathbf{C_{2v}}$ and $\mathbf{C_{3v}}$, you probably said $C_4 + 4\sigma_v$, and this is correct.

There is a whole set of groups $\mathbf{C_{nv}}$ whose members contain the symmetry elements $C_n + n\sigma_v$.

We can even have the group for which n is infinite, namely $\mathbf{C_{\infty v}}$. This group is important because it contains all diatomic molecules in which the two atoms are different (heteronuclear diatomic molecules), such as HF. It also contains all linear molecules that do not have a centre of symmetry, for example HCN.

Are there any other sets of groups?

⬤ What are the symmetry elements of BF_3?

⬤ $C_3 + 3C_2 + 3\sigma_v + \sigma_h$.

The axis of highest symmetry is the C_3 axis, but BF_3 does not belong to the group C_{3v} because it contains more symmetry elements than the molecules in that group.

⬤ What are the extra symmetry elements in BF_3?

⬤ $3C_2 + \sigma_h$.

The groups to which BF_3 belongs is labelled $\mathbf{D}_{3h}$. D_3 is used rather than C_3 for a threefold axis when a molecule has three twofold axes at right-angles to the threefold axis. The subscript h stands for horizontal, and is added if there is a horizontal plane of symmetry.

The symmetry elements of the group $\mathbf{D}_{3h}$ are therefore $C_3 + 3\sigma_v + 3C_2 + \sigma_h$.

There is another set of groups $\mathbf{D}_{nh}$, which contain the elements $C_n + n\sigma_v + nC_2 + \sigma_h$. If n is an even number or infinity, the molecule also possesses a centre of symmetry, i.

The group $\mathbf{D}_{\infty h}$ contains all diatomic molecules with two identical atoms, for example F_2 and N_2, and all linear molecules with a centre of symmetry, such as ethyne, $HC\equiv CH$.

The two sets of groups $\mathbf{C}_{nv}$ and $\mathbf{D}_{nh}$ are among the most important for chemists, because molecules belonging to them are quite common.

There are two other important groups that we have not yet mentioned but which are also common. These are the groups of objects with the same symmetry elements as a tetrahedron and as an octahedron. Molecules and other objects with the same symmetry elements as a tetrahedron, for example CH_4 and SiF_4, are labelled $\mathbf{T}_d$ (T for tetrahedron and d for dihedral*). Octahedral molecules, for example SF_6, belong to the symmetry point group $\mathbf{O}_h$ (O for octahedron and h for horizontal). We are not going to look at these groups in detail. They contain a large number of symmetry elements. It is usually very easy to see whether a molecule is octahedral or tetrahedral, but note that only regular tetrahedral and octahedral molecules belong to $\mathbf{T}_d$ and $\mathbf{O}_h$, respectively. Thus CH_4 belongs to $\mathbf{T}_d$ but CH_3F does not.

When using VSEPR to predict the shape of a molecule, we often refer to a molecule with four repulsion axes as adopting a tetrahedral arrangement, even if the four repulsion axes are not identical. To belong to $\mathbf{T}_d$, however, the molecule or ion must have four *identical bonds* arranged tetrahedrally. Many metal complexes are tetrahedral or octahedral, and if you go on to study the chemistry of the transition elements, you will find these two groups play an important role.

There is one more family of groups that we want to introduce here, and that is the set $\mathbf{C}_{nh}$. These possess a $\mathbf{C}_n$ axis, a horizontal plane of symmetry and, if n is even, a centre of symmetry. There are a fair number of molecules belonging to $\mathbf{C}_{2h}$, for example *trans*-1,2-difluoroethene, but molecules belonging to $\mathbf{C}_{nh}$ with n greater than 2 are relatively rare.

---

* Dihedral symmetry planes will not be considered further. Molecules in the group $\mathbf{T}_d$ do not have a unique principal axis but have several axes of the same order. The σ_d planes each contain one of these axes but neither contain the other axes nor are at right-angles to them.

Finally there are some groups for molecules that contain few symmetry elements. For example, HOF has only one symmetry element, a plane of symmetry. It belongs to the group C_s. Molecules with only an axis of symmetry belong to a group C_n, where the axis is n-fold. Molecules whose only symmetry element is a centre of symmetry belong to the group C_i and those with no symmetry elements belong to C_1.

By looking at the key symmetry elements of a molecule, we can assign it to a symmetry point group. We have written the rules as a flow chart. This is printed on the fold-out page at back of the Book. To find the point group of a particular molecule, start at the top of the page and answer the questions as you work down. Let's see how this chart works with a molecule whose symmetry point group we know, BF_3.

- First we have to decide if BF_3 is tetrahedral.
- The answer to this is no, so we follow the no line to the next question.

- Is BF_3 octahedral?
- No.

- Has BF_3 a C_n axis?
- Yes.

- What is the largest value of n?
- Three. This is the value of n you must now use in all subsequent questions in the chart for this molecule.

- Are there three C_2 axes perpendicular to the C_3 axis?
- Yes.

- Is there a plane of symmetry perpendicular to the C_3 axis?
- Yes.

So our chart tells us that BF_3 belongs to the point group D_{3h}, which is correct. Let's use the flow chart to find the symmetry point group of another molecule you have met.

Let's try benzene, C_6H_6. Benzene is neither tetrahedral nor octahedral.

- Does it have any axes of symmetry?
- Yes. It has one sixfold axis and six twofold axes.

- The largest value of n is six. Does benzene have six C_2 axes perpendicular to the C_6 axis?
- Yes. The six C_2 axes are along the C—H bonds and half-way along the C—C bonds.

- Is there a plane of symmetry perpendicular to the C_6 axis?
- Yes: the plane of the molecule.

So benzene belongs to D_{6h}.

If you find it easier to compare molecules visually rather than use a flowchart, you may find Figure 6.17 helpful. This shows some shapes having the symmetry properties of their point group. The shapes for C_{nv} are pyramids, whereas those for D_{nh} are cylinders, indicating that the molecule is different or the same at either end of the principal axis. To find the point group of your molecule, select the shape that looks closest and read off the point group. Note that only symmetry point groups C_{nv}, C_{nh} and D_{nh} are shown. Molecules belonging to T_d and O_h are easy to spot. For molecules with one or no symmetry elements, you will have to assign the point group from the symmetry element.

Many molecular orbital programs determine the symmetry point group of a molecule before using this information to simplify the calculations, but some require you to state the point group in the information that you give the computer.

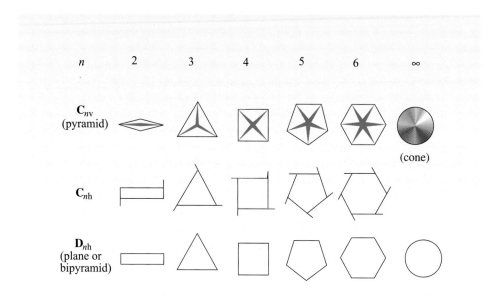

Figure 6.17
Figure for allocating symmetry point groups.

6.7 Summary of Section 6.6

1 All objects with the same symmetry elements belong to the same symmetry group.

2 By following the rules in the flow chart or by comparing your molecule to the shapes in Figure 6.17, you can assign a given molecule to its symmetry point group.

3 The most important point groups for chemists are C_{nv}, D_{nh}, T_d and O_h.

QUESTION 6.9

Use your flow chart or Figure 6.17 to determine the symmetry point groups of the following molecules: (a) SO_2 (V-shaped); (b) carbon monoxide, CO; (c) methanal, $H_2C=O$.

QUESTION 6.10

Below are two lists. The first contains a number of molecules and the second some symmetry point groups. Match each molecule in the first list with a point group in the second list.

Molecule	Symmetry point group
(a) H_2	(i) $\mathbf{C_{2v}}$
(b) ethene	(ii) $\mathbf{C_{3v}}$
(c) trifluoromethane, CHF_3	(iii) $\mathbf{C_{\infty v}}$
(d) O_3 (bent)	(iv) $\mathbf{D_{2h}}$
(e) HCl	(v) $\mathbf{D_{\infty h}}$

QUESTION 6.11

Which of the following symmetry point groups contain molecules that have a centre of symmetry: $\mathbf{D_{2h}}$, $\mathbf{C_{5v}}$, $\mathbf{C_i}$, $\mathbf{D_{3h}}$, $\mathbf{C_{6v}}$?

QUESTION 6.12

Show that the CF_4, CF_3Cl and CF_2Cl_2 molecules, although all based on a tetrahedral shape according to VSEPR theory, belong to different symmetry point groups by considering the number of threefold axes in each molecule.

6.8 Symmetry and orbitals

We said at the beginning of this Section that symmetry was used to label molecular orbitals. You saw something of this with diatomic molecules. Diatomic molecules belong to $\mathbf{C_{\infty v}}$ or $\mathbf{D_{\infty h}}$. σ orbitals are unchanged by all the symmetry operations of $\mathbf{C_{\infty v}}$, whereas σ_g orbitals are unchanged by all the symmetry operations of $\mathbf{D_{\infty h}}$. For π orbitals, however, reflection through one of the vertical planes of symmetry causes a change of sign (Figure 6.18).

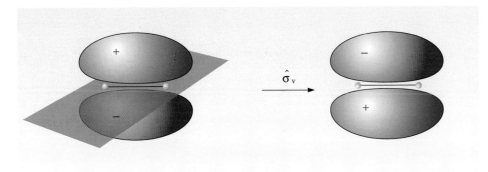

Figure 6.18
Reflection of a π orbital through σ_v.

In $\mathbf{D_{\infty h}}$ there is a centre of symmetry. You saw that orbitals that remained unchanged after inversion were labelled g and those that changed sign were labelled u. Linear molecules with more than two atoms also belong to $\mathbf{C_{\infty v}}$ or $\mathbf{D_{\infty h}}$ and their orbitals are labelled similarly. Ethyne, C_2H_2, for example has σ_g and π_u bonding orbitals because it belongs to $\mathbf{D_{\infty h}}$.

Similar labelling rules apply to molecules in other groups. We will not go into details but note that, for non-linear molecules, molecular orbitals which are unchanged by all operations of a symmetry point group are labelled 'a' with a subscript '1' if the molecules has an axis of symmetry, a subscript 'g' if the molecule has a centre of symmetry and a superscript ′ if the molecule has a plane but no axis of symmetry. For example the bonding molecular orbital for water shown in Figure 5.12a, reproduced in Figure 6.19, is labelled a_1.

Figure 6.19
a_1 orbital of water.

CALCULATIONS IN PRACTICE

7

You have studied the ideas behind molecular modelling, but how would you actually carry out a calculation? In this Section we present a brief overview of how modelling is done. This is reinforced on the CD-ROM.

Suppose, then, you want to instruct a computer to perform a calculation. The first thing you have to do is describe the molecule or crystal you are interested in. The computer needs to know what type of atoms are present and their position. This information can be typed in, but increasingly programs are linked to graphics programs rather like WebLab ViewerLite™ so that you can draw the molecule. It does not even matter if you are no good at drawing since these programs often have libraries of groups of atoms, e.g. a carbonyl group or a benzene ring which you can use to build up your system.

The program will need to know the symmetry point group of your molecule, but it may be able to work this out from the structure. You may sometimes need to explicitly tell the computer the point group, for example if you want to find the lowest energy for a particular conformation.

Then you will need to decide what type of calculation to do. You may have a large molecule such as a protein and want to choose molecular mechanics. Computer power has been, and is, increasing rapidly so that even quite large molecules such as metal complexes and organic molecules of a few hundred atoms can be treated quantum mechanically. Even with very large systems it is becoming increasingly common to treat part of the molecule such as an active site on an enzyme by quantum mechanics and the rest by molecular mechanics.

If you choose quantum mechanics you then have several possible techniques available. The earliest, successful methods for larger molecules were the **semi-empirical** methods. These replaced some of the terms in the solution of the Schrödinger equation by parameters obtained from experimental measurements. As in molecular mechanics, these parameters were transferable and semi-empirical methods were widely adopted by organic chemists. Even today these are used for very large molecules, for calculations where a quick approximate answer is sufficient and to obtain starting geometries for more accurate calculations to find the equilibrium geometry.

Ab initio methods attempt to solve the Schrödinger equation exactly, but in practice have to make some approximations. They do not, however, use any parameters obtained from experiment.

A more recent introduction to quantum chemistry is **density-functional methods** (DFT). These in fact pre-date the other methods but were for a long time used primarily by physicists and were applied to solids. The electron density is given by the square of the wavefunction, and is a measure of how much electron there is at any point. Working from the realization that all properties of a molecule in its lowest energy state can be calculated from the electron density, Walter Kohn and his co-workers re-wrote the wave equation in terms of electron density to show that the exact ground-state energy can be expressed as a mathematical function of the electron density. This mathematical function is called the *density functional*. Unfortunately the precise form of the density functional could not be determined. DFT programs therefore use a variety of functionals that have been shown to give accurate results for molecular calculations but are approximations to the true functional. The electron density used in these programs is calculated by assuming that electrons occupy orbitals as in ab initio Hartree–Fock (p. 34) methods.

If you choose molecular mechanics as in Section 2, you may be asked to choose a data set. If you select a semi-empirical, ab initio Hartree–Fock or density-functional method you will need to choose a basis set for each type of atom. Many programs have a library of basis sets built in that you can simply select by name.

Choosing a DFT method will also require you to choose a density functional. On the CD-ROM we have chosen a popular density functional for molecular calculations known as B3LYP.

Finally, you need to tell the computer what properties of the molecule you want calculated. Do you want to know the equilibrium geometry? Are you perhaps interested in the atomic charges or dipole moment? Do you want to calculate spectroscopic properties? Most programs today include a range of properties which you can choose from, and some of these are illustrated on the CD-ROM.

COMPUTER ACTIVITY 7.1

You should now study *Molecular Modelling* on the CD-ROM associated with this Book. If you studied the molecular mechanics section at the end of Section 2, study the quantum chemistry methods section now.

This allows you to choose quantum chemical methods for particular problems and illustrates some of the features of quantum chemistry computations.

In Sections 2–7 we hope we have given you a flavour of the range and power of computational chemistry for molecules. In the last Section we consider solids.

BONDING IN SOLIDS

8

8.1 Metals

We shall not treat metals in any great detail, such studies being of more interest to metallurgists and physicists than to the majority of chemists. Metals in bonding terms are, however, particularly simple and we can use them to set up more general theories.

Some of the characteristic properties of metals are high electrical and thermal conductivity and a shiny appearance (Figure 8.1), and these can be explained using the descriptions we now develop. We start with a very simple model, the free-electron model.

8.1.1 Free electron model

In the simplest form of this we imagine a crystal of metal as a box in which the valence electrons are free to roam. If we take one electron in such a box and calculate the allowed energy levels using quantum mechanics, we obtain a set of levels. The spacing of the levels depends on the size of the box; the larger the box the more closely spaced are the levels. If we take a small box with sides equal to the diameter of a hydrogen atom, 100 pm, the spacing is very close to that between the two lowest energy states of the hydrogen atom.

Figure 8.1
Sodium metal.

⬤ What are the two lowest states of the hydrogen atom?

⬤ Those with $n = 1$ and $n = 2$, that is 1s and 2s/2p (of equal energy).

Although the energy gap between $n = 1$ and $n = 2$ is close to the energy gap between the two lowest states, as n increases the match between experiment and calculation worsens.

⬤ The spacing between the calculated energy levels increases with n. Is this the pattern seen for the hydrogen atom?

⬤ No. The spacing between successive levels of n in the hydrogen atom decreases as n increases.

Such a model is thus only a crude approximation for an atom or molecule but for a metal it illustrates some useful features which apply to more accurate models as well.

If we take a crystal-sized box, say 1 mm, then the spacing is so small that the levels effectively form a continuous band of allowed energy. This formation of **energy bands** is a feature that is found both experimentally and in more accurate calculations for continuous solids.

The wavefunctions for the electron resemble waves; they must be zero at the edges of the box and so must consist of a whole number of half-waves as in Figure 8.2.

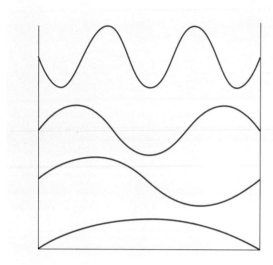

Figure 8.2 Waves in a box.

We can regard these wavefunctions as orbitals for a solid. As in molecules, electrons occupy these orbitals, two of opposite spin in each one. For a metal when all the available electrons are assigned to orbitals, the energy band is only partially filled. This can be represented as in Figure 8.3.

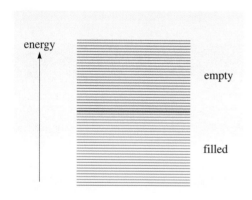

energy

empty

filled

Figure 8.3 Partially filled energy band.

In the presence of an electric field, the energy levels at the negative end of the metal crystal are higher in energy than those at the positive end (Figure 8.4). Electrons can now move from energy levels at the negative end to unoccupied levels at the positive end which are now lower in energy. This process transports electrons through the crystal; in other words an electric current flows.

This is a very simplified model, however, and not much use for other solids. It can be made more realistic by including the ionic cores. These are placed at their crystallographic positions. If we consider magnesium, on the simplest model we

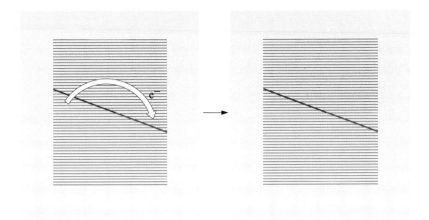

Figure 8.4 Energy levels for a metal in an electric field.

would have two electrons per atom moving freely in the box representing the crystal. Now if we place Mg^{2+} ions at the magnesium atom positions the valence electrons are no longer completely free but have to negotiate round the ions.

Increasingly sophisticated ways of including the ions have been developed and programs based on this idea can reproduce properties of metals, semiconductors and ionic solids quite accurately. We are not going to study this method further, however, but consider an approach that is closer to the way chemists view bonding in molecules.

8.1.2 Molecular orbital theory of solids

In Section 4, we saw how atomic orbitals may be combined to make molecular orbitals. Where a pair of atomic orbitals combines to make molecular orbitals, one bonding and one antibonding combination result; in general, the number of molecular orbitals that can be made equals the number of atomic orbitals which combine. In addition, we know that the average energy of the resulting molecular orbitals is the same as that of the atomic orbitals that were used to generate them. These generalizations are true no matter how many orbitals from however many atoms are combined.

If we start with one atomic orbital on each atom and make a molecule containing n atoms, we form n molecular orbitals. In our theory, a crystal of a metal or one of the other continuous solids is regarded as one molecule.

⬤ Magnesium metal has a density of $1\,740\ kg\ m^{-3}$. A typical metal crystal has a volume of about $10^{-12}\ m^3$ (This is the volume of a cube of side one-tenth of a millimetre.) How many atoms would a typical crystal (or molecule) of magnesium contain?

⬤ The mass of the crystal is $1\,740 \times 10^{-12}\ kg$ or $1.74 \times 10^{-6}\ g$. Because 1 mole of magnesium has a mass of 24.3 g and contains 6.02×10^{23} atoms, one atom

weighs $\dfrac{24.3}{6.02 \times 10^{23}}$ g and 1.74×10^{-6} g contain $\dfrac{1.74 \times 10^{-6} \times 6.02 \times 10^{23}}{24.3}$

atoms or about 4.3×10^{16} atoms.

So if we took one 3s orbital from each magnesium atom in the crystal, we could combine them to make about forty thousand trillion molecular orbitals!

Figure 8.5 shows the energy levels for combinations of up to 20 atoms, arranged in a linear chain. Note that the range of energies occupied by the molecular orbitals does not increase proportionately as the number of atoms increases; the consequence is that the molecular orbitals get progressively closer together as the number of atoms increases. This conclusion can be extended to molecular orbitals formed from a three-dimensional array of atoms. It means that in our crystal, therefore, we have a large number of energy levels, all of about the same energy. The energy differences between these levels are very small.

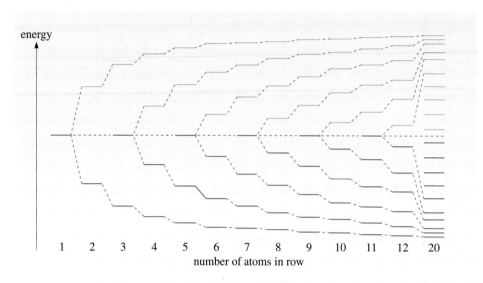

Figure 8.5
The energy levels for a chain of *n* atoms, each atom interacting only with its nearest neighbour. Note that by $n = 20$ the levels are getting very close together, and that the energy difference between the lowest bonding level and the highest antibonding level does not increase indefinitely.

For molecules, atomic orbitals could only combine in certain ways determined by the symmetry of the molecule. For crystals, we combine orbitals in regular repeating patterns which depend on the symmetry of the crystal. For example, Figure 8.6 shows some combinations of eight s orbitals along one axis of a crystal. Note that the sign of the orbitals changes according to a regularly varying pattern e.g. $+ - + - + - + -$ or $++ -- ++ --$. Each pattern has associated with it a quantum number, k.

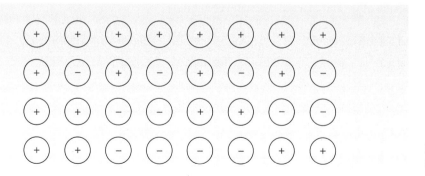

Figure 8.6
Combinations of eight s orbitals along one axis in a solid.

The energies of the occupied molecular orbitals formed from the 3s orbitals on each atom in sodium metal are spread over a range of 4.8×10^{-19} J. Comparison of the densities show that sodium crystals contain about half as many atoms as magnesium crystals. Assuming that the molecular orbitals are evenly spaced in energy, what is the energy difference between two neighbouring levels?

● $(4.8 \times 10^{-19}/2.15 \times 10^{16})$ J, or about 2×10^{-35} J.

This is an extremely small amount of energy. It is about one twenty-thousand-trillionth of the energy of one photon of visible light. It is so small that, as in the free electron model, we can regard the energy levels as forming a continuous range of energies, called an energy band, in this case a 3s energy band. So we can describe the bonding in metals by regarding a crystal as a molecule and combining atomic orbitals on every atom in the crystal.

Band theory, however, applies to any continuous solid, not just metals, so that in this theory certain non-metallic solids, for example diamond and solid noble gases, are also described as having electrons in molecular orbitals that cover the entire crystal. What is it that distinguishes these solids from metals?

Metals are usually distinguished by their high electric and thermal conductivities and their shiny appearance (metallic lustre). So we have to see what it is about metals that makes heat and electricity move through them fairly easily.

The free electron model attributes the high conductivity of metals to the fact that electrons are delocalized throughout a crystal and that there are unoccupied levels in the energy band. In band theory we have molecular orbitals that cover the entire crystal, and so it seems possible that the electrons in them may be free to move through the crystal.

If we compare the number of electrons in metallic solids with the number in non-metallic solids, we find that in the non-metals there are just enough electrons to completely fill all the energy levels in the energy bands that are occupied. The metals have enough to only partly fill the energy levels in the highest energy band that is occupied.

Let's take lithium as an example. Atomic lithium has the electronic configuration $1s^2 2s^1$. The valence 2s orbitals can be combined to form orbitals covering the whole crystal, and these orbitals will form an energy band.

● How many energy levels will the 2s energy band in lithium metal contain?

● It will contain n, where n is the number of atoms in the crystal.

● How many electrons are there available to occupy this energy band?

● There are n: each atom makes available its 2s electron.

Each energy level, though, can contain two electrons, and so only half the energy levels in the 2s band in lithium metal are filled (Figure 8.7). This means that there are a large number of unoccupied energy levels available into which electrons can be excited by absorbing a very small amount of energy. When the excited electron returns to its ground state, the energy lost may be absorbed by another electron and so can be transferred through the solid by this absorption/emission mechanism. The availability of unoccupied energy levels of closely similar energy is what distinguishes a metallic solid from a non-metallic or semi-metallic one.

As in the free electron theory, when an electric field is applied, the energy levels vary across the solid. Where a partially filled band exists, electrons are easily excited to one of the unoccupied levels of that band in a part of the crystal where the energy of these levels is lower and so can carry energy through the solid.

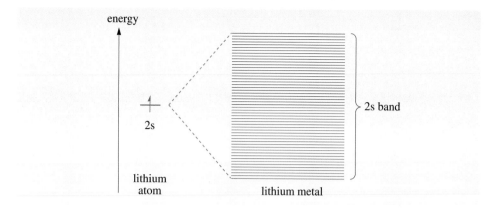

Figure 8.7
Orbital energy-level diagram for lithium metal. Occupied energy levels are indicated by blue lines and unoccupied energy levels by red lines.

Unlike free electron theory, we now have the possibility of completely filled bands and it is the absence of partially filled bands that distinguishes non-metals and semi-metals. An electron in a full band cannot be excited to a higher energy level in the same easy way as in a partially filled band; sufficient energy must be available to enable the electron to reach a level in the next highest energy band.

Thus, in metals that have a partially filled band, electric current is easily transferred through the solid. In non-metals it is necessary to apply a large amount of electrical energy before the electrons can transfer energy.

8.2 Semiconductors

Metals are good conductors of electricity but their conductivity decreases with increasing temperature. This happens because atoms in the lattice through which current-carrying electrons travel are vibrating with an amplitude that increases with rising temperature, and the electron movement is slowed down by interaction with the lattice vibrations.

In the 1830s, however, Michael Faraday noticed that the conductivity of silver sulphide crystals increased as the temperature rose. This led to the identification of a class of substances known as **intrinsic semiconductors**. These substances were distinguished by their relatively low conductivity (roughly in the range 10^{-8}–$10^{1}\,\mathrm{S\,m^{-1}}$ compared with a conductivity of $10^{4}\,\mathrm{S\,m^{-1}}$ for a good metallic conductor or $10^{-14}\,\mathrm{S\,m^{-1}}$ for a good insulator), although there is no sharp distinction between the categories. The characteristic property of a semiconductor is the one noted by Faraday, i.e. that the conductivity rises with temperature (Box 8.1).

BOX 8.1 Conductivity and resistance

Electrical conductivity measures the ease with which electrons pass through a substance. It is measured in the unit $\mathrm{S\,m^{-1}}$. S is the symbol for the unit siemens, which is equivalent to $\mathrm{ohm^{-1}}$. An ohm (Ω) is a measure of electrical resistance, which tells us at what rate electrons will move through a substance for a given applied voltage. An ohm is the resistance of a conductor, which, when a potential difference of 1 volt is applied across it, passes a current of 1 ampere; that is, 1 ohm = 1 volt/1 ampere (in absolute terms defined as $1\,\mathrm{m^2\,kg\,s^{-3}\,A^{-2}}$).

The elements that fall into this category are silicon, germanium, selenium and tellurium. Iodine shows some semiconducting properties, and phosphorus, sulfur and arsenic can each be obtained in a crystalline form that has the properties of a semiconductor, although this is not the most stable form of these elements under normal conditions.

Semiconductors, particularly silicon, are at the heart of electronic devices such as computer chips (Box 8.2). These devices rely on the tailoring of semiconductor properties by the addition of small amounts of impurities. We shall discuss the effect of impurities shortly but first we consider the pure solids.

BOX 8.2 A silicon chip

Silicon chips for use as transistors, solar cells or electronic circuits are carefully built up layer by layer with differing amounts and types of impurities. The amount of impurity needed to produce useful properties is extremely low, 1 part in a million. This is below the normal level of impurity found in silicon so the first step in preparing these devices is to produce ultra pure silicon with an impurity level of less than 1 part in a billion. Crystals of pure silicon are heated at one end until a molten zone is obtained. The heat is then moved across the crystal. At each point impurities concentrate in the melt and are moved along to the next section, finally being removed from the end.

Carefully controlled amounts of impurity are allowed to diffuse into the solid to form the basis of electronic devices (Figure 8.8) .

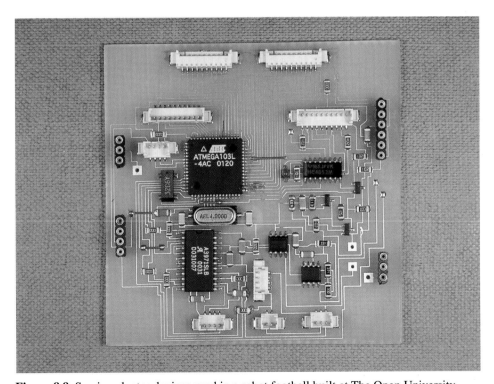

Figure 8.8 Semiconductor devices used in a robot football built at The Open University.

8.2.1 The band structure of diamond and silicon

You saw that lithium formed a band from the 2s orbitals on each lithium in the crystal and that there were insufficient electrons to completely fill this band.

⬤ Consider the bonding in beryllium ($1s^2 2s^2$) on the basis of Figure 8.7. Would you expect beryllium to show metallic properties?

⬤ The two electrons that fill the 2s orbital in a Be atom would also completely fill the 2s band in solid beryllium. No metallic properties would be expected.

However, beryllium is in fact a bright silvery metal with good electrical conductivity.

In the early Groups of the Periodic Table, 2s and 2p orbitals are not well-separated in energy.

As the 2s orbitals combine and spread out to form a band, it will overlap with the band generated by the combination of 2p orbitals. So, in solid beryllium (and similarly in other solid elements in Groups I–III (1, 2 and 13)), the band generated by combination of atomic orbitals is not a pure 2s band, but one continuous band made by combination of both 2s and 2p orbitals. In consequence, there are unoccupied levels available in the energy band of solid beryllium into which electrons may be excited. This allows the development of metallic properties.

Figure 8.9a illustrates the origin of the band derived from the mixing of s and p orbitals.

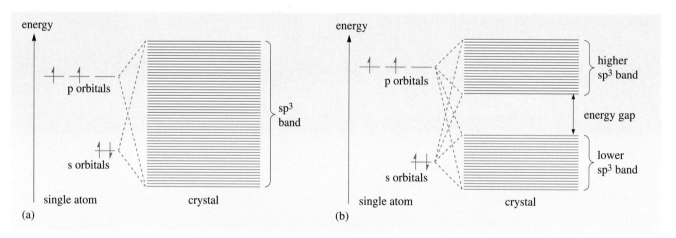

Figure 8.9 Energy bands formed from ns and np atomic orbitals for (a) a high coordination, e.g. eight-coordinate *bcc*, structure and (b) a four-coordinate (tetrahedral) structure, showing filled levels for $4n$ electrons.

⬤ How many electrons are needed to fill this band?

⬤ Eight electrons are needed to fill the four atomic orbitals (one s and three p); in the same way, the *band* will be completely filled when eight electrons per atom are supplied.

We shall now consider what situation will apply for the Group IV (14) elements (outer electronic configuration ns$^2 n$p^2) if overlapping of s and p bands generates one continuous band.

⬤ On this basis, should we expect the elements to have the properties of metals or of insulators?

● We would expect them to be metallic.

In fact, the lighter Group IV (14) elements, carbon, silicon and germanium, are not metals. Carbon, as diamond, is one of the best electrical insulators known. It seems that the 'one continuous band' hypothesis is not a good way of describing energy levels of these elements in the solid state. The band that is formed from the valence electrons of carbon appears to be full with four electrons per atom rather than eight.

The metals we have discussed so far adopt close-packed structures and this provides extensive overlap of the atomic orbitals so that a wide continuous band is formed. In the tetrahedral (diamond) structure (Figure 8.10) adopted by solid carbon, silicon and germanium, the overlap is less, and as a consequence of the crystal symmetry the atoms act as though they were sp^3 hybridized and bonded to their four neighbouring atoms. The s/p band separates into a pair of 'bonding' and 'antibonding' energy bands (Figure 8.9b). Each of these bands can hold four electrons per atom; a total of eight electrons per atom is needed to fill both bands. The extent of splitting of the bands, or the energy gap between the bonding and antibonding combinations has important consequences for the properties of the solid element.

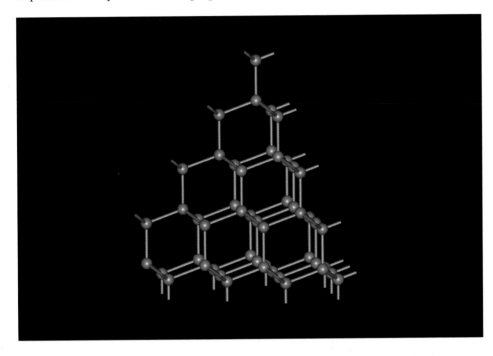

Figure 8.10
Diamond structure.

Band theory explains the properties of insulators by proposing that these have every level in the highest occupied energy band full and so there are no available empty levels into which electrons can be promoted and hence carry energy. At 0 K, this is also true of semiconductors. Increase of conductivity with temperature implies that it is no longer true for semiconductors at higher temperatures; at room temperature, thermal energy is shared between electrons and the vibration of the crystal lattice.

● For a solid with a full energy band and an empty band close above it in energy, what do you expect to happen as the crystal lattice vibration energy is transferred to the electrons?

● Some electrons will be promoted to the empty energy band.

This higher energy band is now partly occupied, and the electrons can transport electrical energy through the solid. The electrons in the lower band will also be part of the electric current because when some of these are excited to the higher band, it leaves empty energy levels in the lower band. The current for a given potential will be proportional to the number of excited electrons plus the number of vacant levels in the lower band; that is, to twice the number of excited electrons.

This is the situation in a semiconductor. The lower, nearly filled band is known as the **valence band** and the higher, almost empty band as the **conduction band** (Figure 8.11). The energy difference between the valence band and the conduction band is known as the **band gap**.

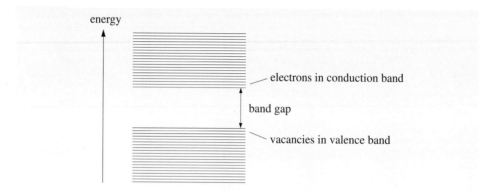

Figure 8.11
Orbital energy-level diagram for a semiconductor at room temperature.

Why does the conductivity of a semiconductor increase with rising temperature?

As the temperature is raised, lattice vibration increases, so more electrons are promoted into the conduction band where they are mobile, and there are more electrons available to carry an electric current. Whereas in an insulator the conduction band is too far above the valence band for an appreciable number of electrons to be promoted to it at normal temperatures, a semiconductor has an empty conduction band lying close above the filled valence band.

Naturally, the smaller the band gap, the more probable is the promotion of electrons. The effect of temperature on conductivity is related to the size of this gap. For an insulator such as diamond, the gap is large and the number of electrons promoted at room temperature is too small to be detected in terms of increased conductivity.

The number of electrons promoted to the conduction band at a temperature T is proportional to $10^{-CE/T}$, where E is the band gap between the valence and conduction bands and C is a constant equal to 3.15×10^{22} J^{-1} K. The band gap for diamond is 1.0×10^{-18} J and for germanium is 1.3×10^{-19} J. Find the ratio of numbers of electrons promoted at room temperature (300 K) for these two solids.

The ratio is equal to:

$$\frac{10^{-(1.3 \times 3.15 \times 10^3)/300} \text{ (Ge)}}{10^{-(1.0 \times 3.15 \times 10^4)/300} \text{ (C)}}$$

$$= \frac{10^{-13.65}}{10^{-105}}$$

$$\approx 2 \times 10^{91}$$

Note how much difference a factor of 10 in the band gap makes. This ratio is so large that we can regard diamond as having no electrons promoted at room temperature, as for any reasonable number of electrons promoted in germanium the fraction promoted for diamond would be so small as to be meaningless.

Germanium, with electrons promoted at room temperature, is a semiconductor, whereas diamond is an electrical insulator.

8.2.2 Compound semiconductors

It is not only elements that can be described by band theory, we can also use it for compounds. Here we consider simple binary compounds (that is, compounds composed of two elements). As in elements, orbitals on each atom overlap, but now we have to consider the relative orbital energies of the two atoms involved. When the elements are very different in electronegativity, for example sodium chloride, then the orbitals of the two atoms are well separated in energy. Consequently the two types of atom form separate bands. In the case of sodium chloride there will be the 3s/3p chlorine band which will be lower in energy than the 3s/3p sodium band. Because the chlorine band is lower in energy, the electrons from the sodium 3s orbitals fill the chlorine band. Thus electrons are transferred from sodium to chlorine giving effectively $Na^+ Cl^-$ as you might expect. Because the chlorine atoms are separated from neighbouring chlorine atoms by sodium atoms, the overlap of the orbitals forming the band is small. The band is narrow (that is, it has a small spread of energies). The narrower a band, the less interaction there is between the atoms; in the extreme case of non-interacting atoms, there would be a set of atomic orbitals (one for each atom) all at the same energy, effectively a band of zero width. Ionic solids such as sodium chloride are insulators, the gap between the chlorine band and the sodium band being large.

When we combine elements that are closer in electronegativity, the bands contain contributions from both atoms. The so-called III–V compounds (which contain one element to the left of the carbon group and one to the right, e.g. gallium arsenide, GaAs) are of particular interest. These compounds adopt four-coordinate structures, e.g. zinc blende or wurtzite, which are similar to diamond but with each type of atom surrounded by four of the other type (Figure 8.12).

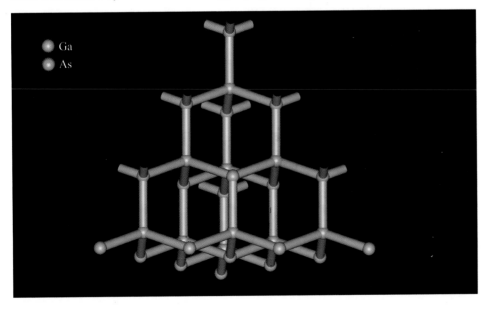

Figure 8.12
Crystal structure of GaAs.

● For a crystal of GaAs containing $2n$ atoms, how many valence electrons are there?

● Each gallium will have 3 and each arsenic 5 valence electrons. There are n atoms of each so there will be $3n + 5n$ electrons or $8n$ in total.

This is the same number as for silicon. As in silicon two s/p bands are formed, each with $4n$ levels.

● Into which band will the $8n$ valence electrons go?

● Into the lower s/p band, which will then be full.

Many of these compounds, like silicon, are intrinsic semiconductors. Because the atomic orbitals from which the bands are formed are not of the same energy for the two types of atom, the lower-energy orbitals in the band contain a greater contribution from one atom and the higher-energy orbitals a greater contribution from the other. This is similar to the situation we found for heteronuclear diatomic molecules such as NO. In the case of gallium arsenide, the lower band will have a greater contribution from arsenic orbitals (Figure 8.13). There is some transfer of electrons from gallium to arsenic as a result but it is less complete than the transfer in ionic solids such as sodium chloride.

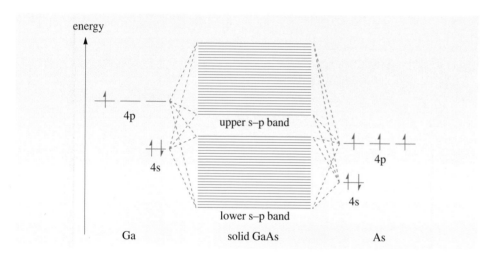

Figure 8.13
Energy-level diagram for GaAs.

Gallium arsenide and other III–V semiconductors are used in some applications in preference to silicon. One such use is in the infrared lasers in CD players.

8.3 Summary of Sections 8.1 and 8.2

1 Metals can be described by a model of free electrons in a box but this model does not explain the properties of non-metals or the change in conductivity with temperature for metals.

2 A modification of the free electron model which includes ion cores has proved useful in computing properties of solids, as has a model in which orbitals for a crystal are built up from atomic orbitals.

3 In a continuous solid we can regard one crystal as a single molecule, and combine atomic orbitals on each atom to form molecular orbitals.

4   The molecular orbitals formed from any one atomic orbital on each atom lie very close together and form a band of energy levels.

5   In metallic solids the valence band (the highest occupied band) is only partially filled.

6   In intrinsic semiconductors the valence band is full but there is a low-lying conduction band into which electrons can be promoted.

7   The lowest-energy empty band in insulators is too far away in energy from the highest-energy occupied band for electrons to be promoted to it at normal temperatures.

8   The electrical conductivity of a metal (that is, the ease with which an electron can move through the metal) decreases with temperature because the mobile electrons are hindered by the increased vibrations of the lattice at higher temperature.

9   The conductivity of a semiconductor increases with temperature because more electrons are promoted into the conduction band as the temperature is raised.

QUESTION 8.1

Consider the following five substances: (a) germanium, (b) calcium, (c) carborundum, SiC, (d) boron nitride, BN, (e) LiAl. Will these solids have metallic, semiconductor or insulator properties if they crystallize with (i) an eight-coordinate *bcc* lattice, (ii) a four-coordinate diamond structure?

QUESTION 8.2

Indium antimonide (InSb) forms crystals with the zinc blende structure (similar to diamond). These crystals are semiconductors. Describe the bonding in a crystal of InSb, and draw an orbital energy-level diagram for the compound. Atomic orbitals for Sb are lower in energy than those for In.

8.4 Photoconductivity

There is more than one mechanism for exciting an electron from its lowest energy level (for example, $n = 1$ in the hydrogen atom) into a higher level. Thermal excitation, the process that gives rise to observation of the Balmer absorption lines (Figure 8.14) in the solar spectrum, is one way, which is very similar to the thermal excitation that we have just been discussing for semiconductors with a small band gap.

⬤   Can you suggest another source of energy that can be used to promote electrons in atoms to higher energy levels?

⬤   You are probably familiar with emission and absorption spectra. Emission spectra correspond to energy-losing transitions of electrons that have been excited, for example thermally or by electric discharge. When they drop from the excited level to a lower one, the energy lost may be given out as photons of distinct energies. If these energies are in the visible region, sharp coloured lines are observed.

Electrons in the lower-energy band of a semiconductor can absorb photons of appropriate energy to allow them to be excited to a level in the higher-energy band. The electrons thus excited are mobile and can carry current; this response to

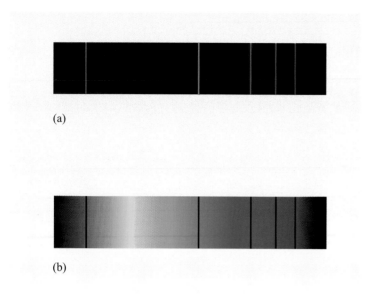

(a)

(b)

Figure 8.14
(a) Emission and (b) absorption spectra
of the Balmer lines of hydrogen.

illumination is known as **photoconductivity**. It has been exploited in devices such as the photocell, which conducts electricity when light is shone on to it but is a poor conductor in the dark. Such devices are used to operate some automatically opening doors, in photocopying machines and in spectrometers. They can be sensitive to visible light or to ultraviolet or infrared radiation. Materials typically used for this purpose include selenium, cadmium sulfide and lead sulfide.

In general, electrons can be excited from near the top of the valence band to near the bottom of the conduction band.

- The energy associated with one photon of visible light ranges from about 2.4×10^{-19} J to about 5×10^{-19} J. The gap between the valence and conduction bands in selenium, a semiconductor often used in photocells, is 2.9×10^{-19} J. Can you suggest why selenium is a photoconductor?

- The energy of virtually any photon of visible light is sufficient to promote an electron from the filled valence band to the empty conduction band.

When light is shone on a photoconducting substance, then, electrons are promoted from the valence band to the conduction band, so that both these bands are now partially full and a current can flow through the substance (Figure 8.15).

- The band gaps in the semiconductors Ge, Si, and GaAs are 1.3×10^{-19} J, 1.9×10^{-19} J and 2.2×10^{-19} J, respectively. How do these energies compare with that of photons of visible light? What is the consequence of illumination?

- Because the photons of the longest-wavelength visible light have an energy of about 2.4×10^{-19} J, all photons of visible light have more than enough energy to promote electrons from the valence to the conduction band in all three materials. Both bands become partially full on illumination, therefore all the electrons in the conduction band (together with some in the valence band) are mobile and are able to migrate toward a positive charge, thus carrying a current.

In the questions in this Book and its associated CD-ROM, you can assume that electrons can be promoted from the top of the valence band to the bottom of the conduction band. Although this is true for many solids, e.g. gallium arsenide, it is not strictly true for others. When an electron is promoted from the valence band to the conduction band, it must go to a level with the same value of the quantum number k that it had originally. In some solids such as silicon this means that an electron at the top of the valence band must absorb light of greater energy than the band gap in order to be promoted to the first allowed level in the conduction band. Conversely, an electron in the bottom level of the conduction band cannot return to the top of the valence band by emitting light. This latter process is important for laser action and is the reason why gallium arsenide is preferred over silicon as a laser material.

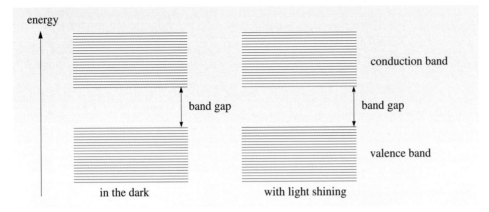

Figure 8.15
The effect of light on the occupancy of the energy bands of a photo-conducting material. We have shown the valence band full in the absence of light. Photoconductors of sufficiently small band gap would have some electrons already in the conduction band at room temperature, but more would be excited to this band when light was shone on to the material.

QUESTION 8.3

The energy of photons of visible light covers the range from about 5×10^{-19} J to about 2.4×10^{-19} J. The band gaps of several semiconductors and insulators are given below. Which of these substances would be photoconductors (a) over the entire visible region, (b) in the ultraviolet only?

Substance	ZnO	TiO$_2$	I$_2$	ZnS	GaSb	CdS
Band gap/10^{-19} J	5.4	4.9	2.1	5.8	1.3	3.8

8.5 Impurity semiconductors

As we mentioned earlier, most semiconductor devices depend on the effects of impurities. Let us see what happens when impurities are added to a semiconductor such as silicon. First we shall see what happens if we add extra electrons.

As the extra electrons cannot be accommodated in the valence band, they will end up in the conduction band. To take an example, suppose we replace a small proportion of the silicon atoms in a crystal with phosphorus atoms. The silicon is said to be *doped* with phosphorus; the phosphorus atoms occupy sites in the lattice that would normally be occupied by silicon atoms. The amount of phosphorus needed is very small; a few parts per million is sufficient and this doesn't change the underlying lattice. Large amounts of phosphorus would disrupt the crystal structure.

Phosphorus has one more electron than silicon. Which orbitals do the extra electrons go into?

Silicon has a filled band of electrons (the lower sp³ (valence) band), so the extra electrons will have to go into the first empty band.

Strictly speaking, the electrons do not enter the empty silicon conduction band, but are in energy levels very close to this band and just below it. Such levels are known as **impurity levels** (Figure 8.16).

The electrons in the impurity levels are very easily promoted into the conduction band of silicon and, as the gap between the two is small enough to be bridged by thermal excitation at normal temperature, they may be considered to merge. Therefore we will not represent impurity levels separately in succeeding diagrams.

What effect will the presence of phosphorus impurity have on the conductivity of silicon at any given temperature?

The conductivity will increase because the extra electrons are very easily promoted to the conduction band.

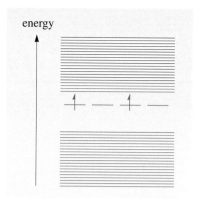

Figure 8.16
Orbital energy-level diagram for an impurity semiconductor: silicon doped with phosphorus.

Silicon with some added phosphorus is an example of an **n-type semiconductor**, because there are extra **n**egatively-charged electrons in the bands.

An n-type semiconductor contains an excess of electrons from the point of view of *bonding* but not, of course, from the point of view of nuclei; in terms of charge, the extra electrons in silicon doped with phosphorus are balanced by the extra positive charge on the phosphorus nuclei. It is important to remember that n-type *semiconductors are electrically neutral*. There is no net charge on silicon doped with phosphorus, for example, because it is composed of neutral atoms.

There is a second type of impurity semiconductor, which has *fewer* electrons than necessary to fill the valence band. We shall call this type of semiconductor **p-type**.

Can you suggest an element that might be doped into silicon to make a p-type semiconductor?

Group III elements have only three valence electrons, so if a small proportion of a Group III element can be mixed into silicon without changing the structure, the result is a p-type semiconductor. Boron is the element most frequently used for this purpose.

Like the n-type semiconductors, silicon doped with boron is composed of neutral atoms and is uncharged. For each boron atom in the crystal there will be one electron fewer in the highest filled band of silicon. Because the valence band is now only partly full, boron-doped silicon is a better conductor than pure silicon. The current in p-type semiconductors is carried by the few mobile electrons in the valence band, and is proportional to the number of empty levels in this band.

We have seen that when intrinsic semiconductors are heated or illuminated, electrons are promoted from the valence band to the conduction band. This generates vacancies (which also permit electron movement) in the valence band. In n-type semiconductors, we feed electrons into the conduction band without affecting the valence band, and in p-type semiconductors we introduce vacancies in the valence band but leave the conduction band empty. The three types of semiconductor are shown in Figure 8.17.

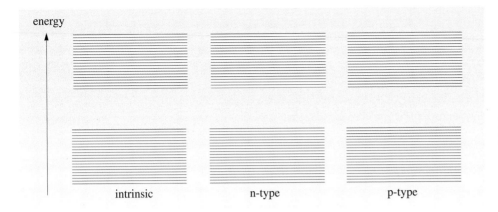

Figure 8.17
Intrinsic, n-type and p-type semiconductors, showing the formation of carriers (electrons in the conduction band and vacancies in the valence band.)

Impurity semiconductors share with intrinsic semiconductors the property of increased conductivity under conditions of illumination or higher temperature. Electrons are easily promoted, thermally or upon absorption of a suitable photon, from the valence to the conduction band, but unless the promoted electron is removed by reason of its migration towards a positive charge (as in an electric field), promotion is rapidly followed by recombination of the electron with a nucleus (Figure 8.18) so the number of electrons in valence and conduction bands remain unaltered.

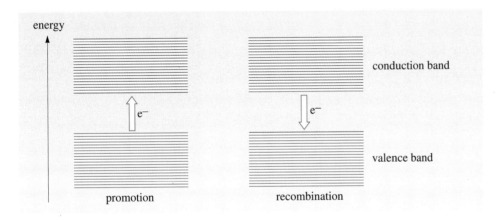

Figure 8.18
The promotion and recombination processes.

Impurity semiconductors are the basis of many electronic devices from transistors to solar cells.

QUESTION 8.4

Which of the following impurities lead to generation of n-type and which to p-type semiconductors: (a) arsenic in germanium; (b) gallium in silicon; (c) germanium in silicon; (d) indium in germanium?

QUESTION 8.5

GaAs, as you have seen, is a semiconductor. Describe the effect on the conductivity of the material of replacing a small proportion of the Ga atoms by (a) Sn atoms and (b) Zn atoms. Illustrate your answer with reference to the occupancy of the conduction and valence bands.

8.6 Summary of Sections 8.4 and 8.5

1　The three types of semiconductor are intrinsic semiconductors, p-type semiconductors and n-type semiconductors.

2　Intrinsic semiconductors have a filled valence band and an empty conduction band. Electrons can be promoted from the valence band to the conduction band by a suitable energy source, such as heat or light, and the semiconductor will then conduct electricity.

3　p-Type semiconductors are intrinsic semiconductors with some atoms replaced by atoms with fewer electrons, and hence they have unfilled levels in the valence band.

4　n-Type semiconductors have some atoms replaced by atoms with more valence electrons, and the extra electrons are effectively in the conduction band.

COMPUTER ACTIVITY 8.1

Now would be a good time to study the *Electrons in Solids* section on the CD-ROM associated with this Book.

8.7 Blue diamonds

The appeal of diamonds lies in their sparkle, due to their high refractive index.

Most diamonds are colourless but diamonds that are yellow, brown, blue, red, pink, purple and orange are found.* One of the most famous coloured diamonds is the blue Hope diamond (Figure 8.19).

● Why are most diamonds colourless?

● Diamond is an insulator and has a large band gap. For an electron to be excited across this gap it has to absorb radiation of higher energy than visible light.

Blue diamonds contain boron as an impurity. Boron has one less valence electron than carbon.

● How is doping with boron going to affect the band structure of diamond?

● The situation here is the same as that in p-type semiconductors. There will be vacancies in the valence band.

Visible light can now be absorbed by electrons from lower down the valence band being excited into the vacant levels. Most of the absorption occurs in the infrared, and the red and yellow regions of the visible spectrum.

● How does absorption of light from the red and yellow regions of the spectrum produce a blue colour?

● We see the light that is transmitted by the diamond. Since the red and yellow regions are absorbed we observe blue light.

QUESTION 8.6

'Canaries' are bright yellow diamonds. Their colour arises due to the presence of nitrogen as an impurity.

(a) What effect will nitrogen have on the band structure?

(b) What part of the spectrum of visible light must be absorbed by the diamonds to cause them to be yellow?

Figure 8.19
A necklace containing the Hope Diamond. The Hope diamond has been linked with ill fortune including the maiden voyage of the Titanic. The dark blue diamond was, however, already in America at the time of that voyage. It appeared mysteriously in London in the early nineteenth century and is believed to have been recut from a large diamond (13.8 g) stolen from a museum housing the French crown jewels in 1792. Its last owner, Henry Winston, donated it to the Smithsonian Institution in Washington, D.C. with the wish that it formed part of a collection to rival that in the Tower of London.

* Diamonds and their structure are discussed in *The Third Dimension*.[2]

LEARNING OUTCOMES

Now that you have completed *Molecular Modelling and Bonding*, you should be able to do the following things:

1   Recognize valid definitions of, and use in a correct context, the terms, concepts and principles in the following Table. (All Questions)

List of scientific terms, concepts and principles introduced in *Molecular Modelling and Bonding*

Term	Page number	Term	Page number
ab initio	91	inversion through centre of symmetry	44
antibonding orbital	42	isoelectronic molecules	64
atomic orbital	28	minimization	18
axis of symmetry	77	molecular mechanics	11
band gap	102	molecular orbital	38
basis set	23	molecular orbital energy-level diagram	42
bond order	53	n-type semiconductor	108
bonding orbital	39	nodal plane	41
Born–Oppenheimer approximation	38	non-bonding orbital	73
boundary surface	24	orbitals	23
centre of symmetry	44	out of phase	25
conduction band	102	π orbitals	49
conjugated molecule	69	p-type semiconductor	108
degenerate orbital	50	paramagnetic substance	54
delocalized orbitals	69	partial charges	16
density-functional method	91	photoconductivity	106
diamagnetic substance	54	plane of symmetry	81
electron-deficient molecules	72	principal axis	81
electron density	23	σ orbitals	44
energy bands	93	semiconductor	98
free electron model	93	semi-empirical method	91
Gaussian function	33	symmetry elements	83
Hartree–Fock method	34	symmetry operations	76
homonuclear diatomic molecule	54	symmetry point group	85
horizontal plane	81	torsion angle	17
hybrid orbital	34	training set	18
impurity levels	108	valence band	102
in phase	25	vertical planes	81
intrinsic semiconductor	98	wavefunction	22

2 Cite the types of forces used to describe solids and organic molecules in molecular mechanics and recognize that these can be used to calculate geometries and other properties. (Questions 2.1–2.4)

3 Understand that different data sets representing these forces may give different geometries and properties. (Question 2.5, Computer Activity)

4 Draw rough sketches of representations of atomic orbitals for 1s, 2s, 2p and 3d electrons and of molecular orbitals formed from these. (Questions 4.8 and 5.1)

5 Describe the effect of (i) increasing atomic number and (ii) increasing principal quantum number on atomic orbitals. (Question 3.1)

6 State whether a given atom in a particular molecule is best described as sp^3, sp^2 or sp hybridized. (Question 3.2)

7 Draw simple molecular orbital energy-level diagrams for diatomic molecules and simple polyatomic molecules. (Questions 4.1, 4.2, 4.4, 4.5, 4.7–4.9 and 5.4)

8 Use orbital energy-level diagrams for simple molecules to predict properties of these molecules such as stability relative to the separate atoms, bond order, paramagnetism. (Questions 4.3, 4.5, 4.6, 4.7, 4.8 and 5.4)

9 Recognize whether a given unsaturated organic molecule will have delocalized orbitals. (Questions 5.1 and 5.2)

10 Recognize the presence of planes, axes and centres of symmetry in a given molecule or object. (Questions 6.1–6.3 and 6.5–6.8)

11 Determine whether or not a molecule could be chiral from its symmetry elements. (Question 6.4)

12 Given a flow chart such as the one on the fold-out page at the end of this Book and/or Figure 6.17, assign a molecule or other object to its symmetry point group. (Questions 6.9–6.12, Computer Activity)

13 Describe and sketch orbital energy-level diagrams for insulators, semiconductors and conductors in terms of energy bands. (Questions 8.1 and 8.2)

14 Describe the origin of photoconductivity and recognize whether a solid with a given band gap will be a photoconductor for a particular range of the electromagnetic spectrum. (Questions 8.3 and 8.6)

15 Describe the effect of given impurities on semiconductors. (Questions 8.4–8.6)

16 Select suitable methods (e.g. ab initio), size of basis sets and type of run (e.g. geometry optimization) for a given calculation. (Computer Activity)

QUESTIONS: ANSWERS AND COMMENTS

QUESTION 2.1 (*Learning Outcome 2*)

Electrostatic forces between Cs^+ and F^-, Cs^+ and Cs^+, and F^- and F^-; intermolecular forces between Cs^+ and F^-, Cs^+ and Cs^+, and $F^-$and F^-; spring forces connecting core and shell for Cs^+ and F^-.

QUESTION 2.2 (*Learning Outcome 2*)

Assuming the important term is the internal energy of the defect, the halide with the lowest Schottky defect energy will have the most defects — that is, KI. The salts all have the same structure, and so the entropy terms would be similar.

QUESTION 2.3 (*Learning Outcome 2*)

The carbon atoms in the methyl groups, CH_3, are attached to one carbon and three hydrogens and so have a partial charge of $3 \times 0.053 + 0 = + 0.159$. The carbon in the CH_2 group is attached to two carbons and two hydrogens, and so has a partial charge of $2 \times 0.053 + 2 \times 0 = +0.106$. The CH carbon has a partial charge of $+0.053$. Finally, there is one carbon (C-2) attached only to carbon and this will have a partial charge of 0.

QUESTION 2.4 (*Learning Outcome 2*)

The different bond terms in ecgonine are $C-C$, $C-H$, $C-N$, $C-O$, $C=O$, $O-H$.

QUESTION 2.5 (*Learning Outcome 3*)

The two sets employ different parameters for the forces, and the geometry at equilibrium is a balance of these forces.

QUESTION 3.1 (*Learning Outcome 5*)

The nuclear charge of sodium is greater than that of lithium, and so the 1s orbital will be concentrated closer to the nucleus in sodium.

The 2s in lithium and the 3s in sodium are valence orbitals. Orbitals of higher n are further out from the nucleus and so the 3s will be concentrated further out. In addition, the 2s has two shells of opposite sign, whereas the 3s has three shells of alternating sign (Figure 3.14).

QUESTION 3.2 (*Learning Outcome 6*)

(i) HCN has a triple bond between C and N as in ethyne, the carbon will be sp hybridized. (ii) The carbon of the CH_3 group will be sp^3 hybridized and that of the COOH group sp^2 hybridized. (iii) Both carbons are sp^3 hybridized. (iv) Both carbons are sp^2 hybridized. (v) The carbons of the CH_3 and CH_2 groups are linked by four single bonds each and are sp^3 hybridized. The carbon of the $C=N$ group is sp^2 hybridized.

QUESTION 4.1 (*Learning Outcome 7*)

You should have obtained a diagram like Figure Q.1 for He_2. There are two electrons in bonding orbitals and two in antibonding orbitals. The electrons in the antibonding orbital will be higher in energy than they were in the 1s, but those in the bonding orbital will be lower in energy, so that overall the electrons in the He_2 molecule will not have a lower electron energy than they did in the two He atoms. Consequently, we would expect the nuclear repulsion and entropy to dominate and He atoms to be favoured over He_2.

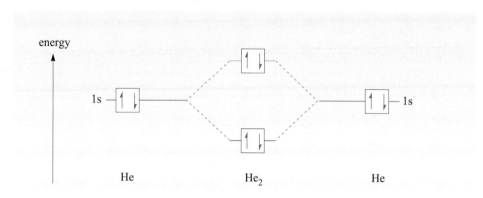

Figure Q.1
Orbital energy-level diagram for He_2.

QUESTION 4.2 (*Learning Outcome 7*)

Be_2 will have eight electrons, since Be has four $(1s^2 2s^2)$. The orbital energy-level diagram for Be_2 is given in Figure Q.2. The electrons are fed into the orbitals, two into the $1\sigma_g$, two into the $1\sigma_u$, two into the $2\sigma_g$ and two into the $2\sigma_u$.

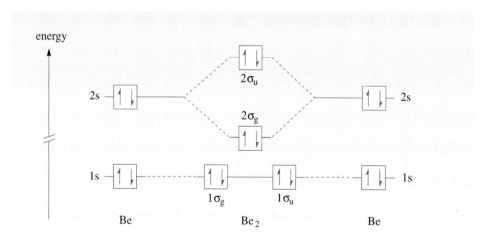

Figure Q.2
Orbital energy-level diagram for Be_2.

You probably predicted that Be_2 would not exist. At room temperature and atmospheric pressure, beryllium is a highly poisonous grey metal. Be_2 is not found at room temperature, and the observed spectrum was obtained by trapping Be atoms in a solid matrix at low temperature. Where pairs of Be atoms were trapped together, they gave spectra characteristic of Be_2. It is probably more stable than predicted from Figure Q.2 because there is some involvement of 2p orbitals in the bonding.

QUESTION 4.3 (*Learning Outcome 8*)

(a) 1; (b) 0; (c) 1. H_2 has one pair of electrons in a bonding orbital and none in an antibonding orbital. He_2 has one pair in bonding and one in antibonding, and Li_2 has two pairs in bonding and one pair in antibonding orbitals. Note that molecules with bond orders of 0, like He_2, are those that we predict not to exist.

QUESTION 4.4 (*Learning Outcome 7*)

Figure Q.3 shows the orbital energy-level diagram for Ne_2. 1s and 2s atomic orbitals and the molecular orbitals formed from them are not shown — these orbitals are full. It has five pairs of electrons in bonding orbitals and five pairs in antibonding orbitals; its bond order is thus 0. You should have predicted that it would not exist, and indeed it has never been observed.

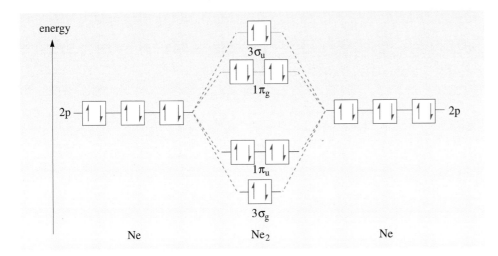

Figure Q.3
Orbital energy-level diagram for Ne_2.

QUESTION 4.5 (*Learning Outcomes 7 and 8*)

The orbital energy-level diagram for O_2^+ is the same as that for O_2 except that there is one less electron to feed in (Figure Q.4). As there is now only one electron in the antibonding orbital $1\pi_g$, the bond order in O_2^+ is $2\frac{1}{2}$ compared with a bond order of 2 in O_2. We would therefore expect (a) the bond length of O_2^+ to be shorter than that in O_2 and (b) the dissociation energy to be larger.

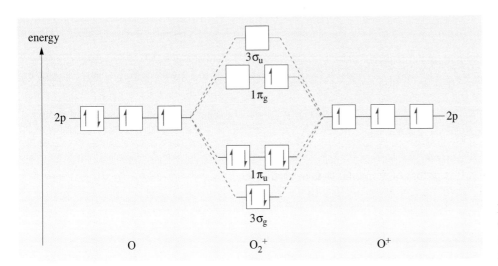

Figure Q.4
Orbital energy-level diagram for O_2^+ showing the 2p orbitals and the molecular orbitals formed from them.

Comment: The observed bond lengths of O_2 and O_2^+ are 121 pm and 112 pm, respectively, and the bond dissociation energies are 498 kJ mol^{-1} and 623 kJ mol^{-1}, respectively. The electron that has to be removed from O_2 to form O_2^+ is antibonding, and we might expect this to be fairly easily removed. This is borne out by the existence of salts of O_2^+, such as $O_2^+[PtF_6]^-$ and $O_2^+[PF_6]^-$. Since O_2 has two electrons in antibonding orbitals, you may have thought that we could also form O_2^{2+} salts, in which the bond order would be 3. The energy required to remove two electrons from O_2, however, is so large that we would need a very large lattice energy to be able to form stable salts.

QUESTION 4.6 *(Learning Outcome 8)*

N_2^+ has one less electron than N_2. The electron lost will come from the $1\pi_u$ bonding orbital, thus reducing the bond order by $\frac{1}{2}$ to $2\frac{1}{2}$. We would therefore expect N_2^+ to have a smaller dissociation energy than N_2.

N_2^- has one more electron than N_2, but because this electron has to go into an antibonding orbital ($1\pi_g$) the bond order of N_2^- is also reduced to $2\frac{1}{2}$. We predict therefore that N_2^+ and N_2^- will have similar dissociation energies, which will be less than that of N_2.

QUESTION 4.7 *(Learning Outcomes 7 and 8)*

The diagram is shown in Figure Q.5.

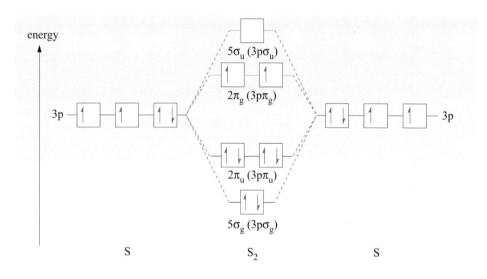

Figure Q.5
The orbital energy-level diagram for S_2 showing 3p atomic orbitals and the molecular orbitals formed from them.

This is almost identical to the diagram for O_2 (Figure 4.23), the only differences being that we use 3p rather than 2p orbitals and that the actual energies of all the orbitals will differ from those in O_2. The bond order is 2 and the molecule is paramagnetic.

QUESTION 4.8 *(Learning Outcomes 4, 7 and 8)*

Figure Q.6 shows an orbital energy-level diagram for OF. This resembles that for NO except that there are two more electrons. These two will go into the $2\pi^*$ antibonding orbital. The highest-occupied molecular orbital is therefore the $2\pi^*$. This will resemble $2\pi^*$ of NO and will have a greater contribution from O 2p than F 2p (Figure Q.7).

OF has a bond order of $1\frac{1}{2}$. OF is thus stable with respect to O and F atoms. It is not observed because it is unstable with respect to other molecules containing O and F,

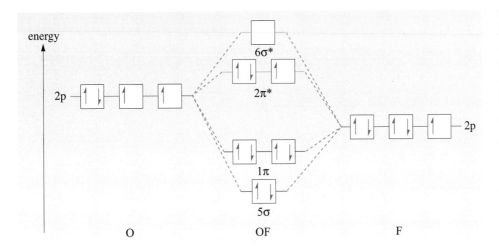

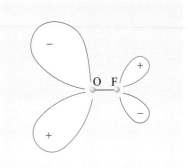

Figure Q.7 $2\pi^*$ orbital in OF.

Figure Q.6 Orbital energy-level diagram for OF.

for example O_2, F_2, OF_2. It is unstable with respect to the disproportion reaction:

$$2OF(g) = OF_2(g) + \tfrac{1}{2}O_2(g)$$

and with respect to the normal forms of O and F, O_2 and F_2:

$$2OF(g) = O_2(g) + F_2(g)$$

QUESTION 4.9 (Learning Outcome 7)

Start by combining C 2p with N 2p and follow the method used for CO. The C 2p and N 2p combine to give 5σ, 1π, $2\pi^*$ and $6\sigma^*$. Since the N 2p lies between the C 2p and C 2s, you now have to mix in the C 2s. Mixing in C 2s will raise the energy of 5σ. The 2s orbitals probably have a greater effect in CN than in CO, but, overall, 5σ is bonding. The orbital energy-level diagram is shown in Figure Q.8.

CN^- has one extra electron which fills 5σ, and so the ion is isoelectronic with CO. CN^- has suitable orbitals to bind to iron and so molecular orbital theory suggests it might if the energies of the CN^- and Fe orbitals are close.

Comment: CN^- is a well-known poison that does in fact bind readily to many metal ions, including iron in haemoglobin.

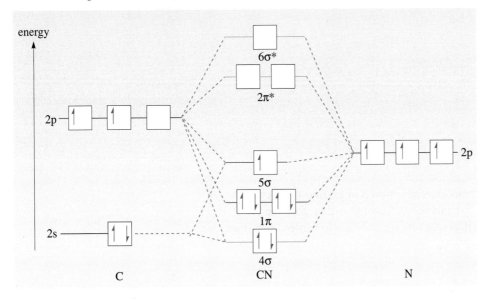

Figure Q.8
Orbital energy-level diagram for CN.

QUESTION 5.1 (*Learning Outcomes 4 and 9*)

Figure Q.9 shows a delocalized π-bonding orbital for cyclohexa-1,3-diene. Note that it involves only four of the six carbons in the ring.

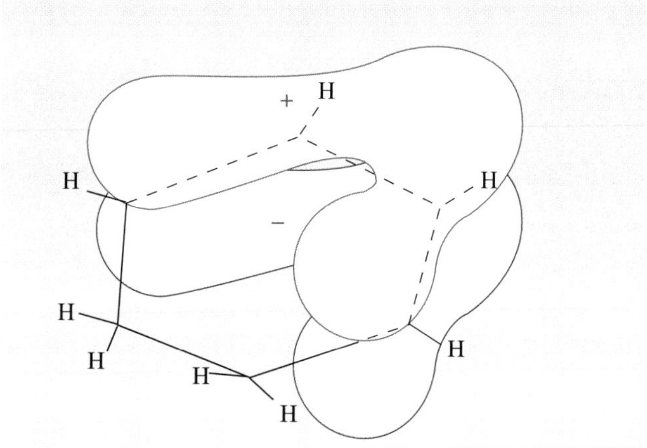

QUESTION 5.2 (*Learning Outcome 9*)

Only hexa-2,4-diene (ii). But-1-ene, (i), has only one double bond. In cyclohexa-1,4-diene there is an sp^3 hybridized carbon separating the two carbon–carbon double bonds. For a delocalized π orbital in alkenes, at least two double bonds are needed and they must be separated by one single bond.

QUESTION 5.3 (*Learning Outcome 8*)

The oxygen 1s and 2s orbitals will not contribute to the bond order. If we use Figure 5.16, there are four electrons (two pairs) in bonding orbitals and none in antibonding orbitals, giving a bond order of 2.

Figure Q.9 Delocalized π-bonding orbital for cyclohexa-1,3-diene.

QUESTION 5.4 (*Learning Outcomes 7, 8*)

Figure Q.10 reproduces Figure 5.17 but with the orbitals labelled as bonding or antibonding. Those orbitals lower in energy than N 2p and H 1s are bonding and those lying higher in energy than N 2p and H 1s antibonding. Note that there are no non-bonding orbitals.

There are six electrons (three pairs) in bonding orbitals and none in antibonding orbitals, giving a bond order of 3.

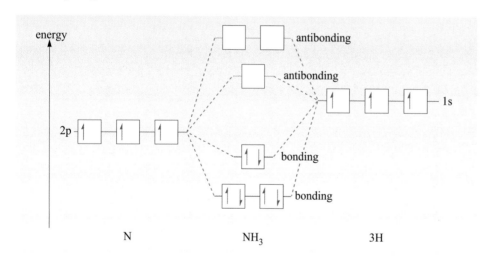

Figure Q.10
Molecular orbital energy-level diagram for NH_3 showing bonding and antibonding orbitals.

QUESTION 6.1 (*Learning Outcome 10*)

(a) BrF_5 is square-based pyramidal. The axial Br—F bond is a fourfold axis and this is the only axis (Figure Q.11).

(b) SO_2 is bent and has a twofold axis like that in H_2O.

(c) $POCl_3$ is pyramidal and the P—O bond forms a threefold axis.

(d) NO_3^- is trigonal planar. There is a threefold axis perpendicular to the plane of the molecule and each N—O bond is a twofold axis.

(e) HBF_2 is planar. The B—H bond is a twofold axis.

QUESTION 6.2 (Learning Outcome 10)

(a) Threefold; (b) sixfold; (c) twofold; (d) fourfold; (e) twofold.

QUESTION 6.3 (Learning Outcome 10)

(i) c; (ii) b; (iii) a; (iv) b; (v) a; (vi) d; (vii) b. The plane of a molecule is always a symmetry plane. In (i) it is neither vertical nor horizontal because HOF has no axis of symmetry. In (ii), the twofold axis is at right-angles to the molecular plane and so the plane is horizontal. In (v), the plane contains the ∞-fold axis and so is vertical. In (vii), there is a threefold axis at right-angles to the plane and three twofold axes in the plane. The threefold axis is the principal axis and so the plane is horizontal. The plane in (iii) contains the P=O bond, which is the principal (threefold) axis and so is vertical. In (iv) the molecular axis of CO_2 is the principal ∞-fold axis and so the plane is horizontal. In (vi) reflection through the plane of the three H atoms in NH_3 would bring the N atom through to the other side, and so this is *not* a plane of symmetry.

QUESTION 6.4 (Learning Outcome 11)

Trialkyl phosphines will be pyramidal like ammonia. For there to be no plane of symmetry, the alkyl groups R^1, R^2 and R^3 must all be different.

QUESTION 6.5 (Learning Outcome 10)

(a) No; (b) no; (c) yes; (d) yes; (e) yes.

QUESTION 6.6 (Learning Outcome 10)

Bent triatomics do not have a centre of symmetry. Only linear triatomics with identical outer atoms will have such a symmetry element. The molecule must therefore be straight: Cl—Ca—Cl.

QUESTION 6.7 (Learning Outcome 10)

The only common geometry for a four-coordinate species that has a centre of symmetry is square planar. Thus $[PtCl_4]^{2-}$ must be square planar.

QUESTION 6.8 (Learning Outcome 10)

(a) BF_3 has four planes of symmetry (one horizontal), one threefold axis and three twofold axes. The symbol for a threefold axis is C_3 and that for a twofold axis is C_2. Our list is thus $C_3 + 3C_2 + 3\sigma_v + \sigma_h$.

(b) SF_4, Structure **Q.1**, has only one twofold axis and two planes of symmetry, both vertical. The list for SF_4 is thus $C_2 + 2\sigma_v$. Note that in SF_4, two sets of fluorine atoms can be distinguished — the axial and the equatorial. These have different S—F bond lengths, and no symmetry operation will move an axial F to the position of an equatorial F or vice versa.

(c) Ethene has three twofold axes, one along the C=C bond and two bisecting this bond at right-angles. It has three planes of symmetry. If we choose one of the twofold axes as the principal axis, then two of these planes are vertical and one is horizontal. Ethene also has a centre of symmetry. The complete list is thus $3C_2 + 2\sigma_v + \sigma_h + i$.

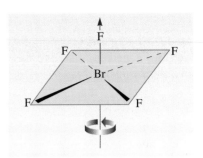

Figure Q.11
The fourfold axis in BrF_5.

Q.1

QUESTION 6.9 (*Learning Outcome 12*)

(a) SO_2 is neither tetrahedral nor octahedral. It has one C_2 axis. The largest value of *n* is thus 2, and there are not two C_2 axes perpendicular to the C_2 axes. There are two planes of symmetry containing the C_2 axis. So SO_2 belongs to the point group $\mathbf{C_{2v}}$. The nearest shape to SO_2 in Figure 6.17 is the top left-hand one.

(b) The molecular axis of CO is a C_∞ axis but there is no C_2 axis perpendicular to it. There are an infinite number of planes of symmetry containing the C_∞ axis. It therefore belongs to $\mathbf{C_{\infty v}}$. You may remember that we said that all heteronuclear diatomic molecules belong to this point group. Looking along the CO axis, the molecule has a circular cross-section but the two atoms differ and so it resembles the circular pyramid, $\mathbf{C_{\infty h}}$.

(c) Methanal is neither octahedral nor tetrahedral. It has a twofold axis of symmetry along the C=O bond. There are no other C_2 axes. There are two vertical planes of symmetry. Methanal belongs to $\mathbf{C_{2v}}$. If you stand methanal with the C=O pointing up, then it resembles the top left-hand diagram in Figure 6.17.

QUESTION 6.10 (*Learning Outcome 12*)

(a) (v): all homonuclear diatomic molecules belong to $\mathbf{D_{\infty h}}$.

(b) (iv): ethene has three C_2 axes, which are all perpendicular to each other, and so can be counted as $C_2 + 2C_2$ perpendicular to the first (hence $\mathbf{D_{2h}}$).

(c) (ii): CHF_3 has a threefold axis along the CH bond and three σ_v planes, each containing the C—H and C—F bond (hence $\mathbf{C_{3v}}$).

(d) (i): no symmetry operation swaps the central atom for one of the outer two, so the fact that the central atom is identical with the outer two atoms makes no difference to the symmetry. O_3 thus belongs to the same point group as SO_2 (hence $\mathbf{C_{2v}}$).

(e) (iii): all heteronuclear diatomic molecules belong to $\mathbf{C_{\infty v}}$.

QUESTION 6.11 (*Learning Outcome 12*)

None of the groups $\mathbf{C_{nv}}$, and only those groups $\mathbf{D_{nh}}$ for which *n* is even, contain a centre of symmetry. $\mathbf{C_i}$ contains a centre of symmetry by definition. So the groups containing molecules with a centre of symmetry are $\mathbf{C_i}$ and $\mathbf{D_{2h}}$.

QUESTION 6.12 (*Learning Outcome 12*)

In CF_4, each C—F bond is a threefold axis so there are four altogether. The C—Cl bond in CF_3Cl is a threefold axis but this is the only one. CF_2Cl_2 has no threefold axes.

Comment: CF_4 belongs to $\mathbf{T_d}$, CF_3Cl to $\mathbf{C_{3v}}$ and CF_2Cl_2 to $\mathbf{C_{2v}}$.

QUESTION 8.1 (*Learning Outcome 13*)

For the eight-coordinate *bcc* lattice (i) there is a continuous sp band which can contain up to $8n$ electrons. For the four-coordinate diamond structure (ii), there are two bands, each of which can hold $4n$ electrons.

(a) Germanium, like carbon and silicon has $4n$ valence electrons. In the *bcc* structure (i), these will partly fill the sp band and the solid will be metallic. In the diamond structure (ii), the lower band will be full and the upper empty. Germanium, like silicon, will be a semiconductor and as it is further down Group IV (14) will have a smaller band gap.

(b) Calcium has $2n$ valence electrons. It is predicted to be a metal with partially-filled bands in either structure.

(c) Carborundum, SiC, is formed from two elements each of which has 4 valence electrons. Like germanium, therefore, it is predicted to be metallic in the *bcc* structure and a semiconductor or insulator in the diamond structure.

(d) Boron has 3 valence electrons and nitrogen 5 so that BN has the same number of valence electrons as diamond. In the *bcc* structure, the band will be partly full and BN is predicted to be metallic. In the diamond structure, the electrons will completely fill the lower band. Boron nitride, like diamond, will be an insulator.

(e) Lithium has 1 valence electron and aluminium 3. LiAl is isoelectronic with Ca and is predicted to be metallic in either structure.

QUESTION 8.2 (*Learning Outcome 13*)

The band structure of InSb will be similar to that of GaAs. Because there are two elements involved, the valence band will contain a greater contribution from Sb, whose orbitals are lower in energy, and the conduction band will have a greater contribution from In. The valence band will be full. This represents a donation of electrons from In to Sb so that this compound can be thought of as having some ionic character $In^{\delta+} Sb^{\delta-}$. The orbital energy-level diagram for InSb is shown in Figure Q.12.

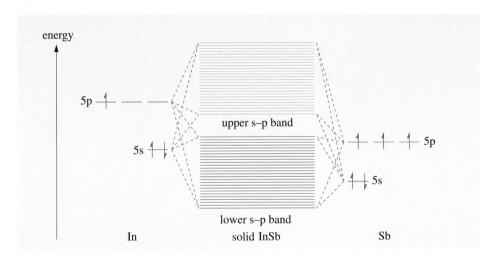

Figure Q.12
Orbital energy-level diagram for InSb.

QUESTION 8.3 (*Learning Outcome 14*)

To be a photoconductor over the entire visible range, the solid must have a band gap of less than 2.4×10^{-19} J. Only I_2 and GaSb fulfil this condition.

To be a photoconductor in the ultraviolet only, the solid must have a band gap greater than the energy of all visible photons, that is greater than 5×10^{-19} J. ZnO and ZnS fulfil this condition.

TiO_2 and CdS will be photoconductors over part of the visible range.

QUESTION 8.4 (*Learning Outcome 15*)

(a) is an n-type semiconductor because arsenic has one more valence electron than germanium. (b) and (d) are p-type semiconductors because indium has one less valence electron than germanium, and gallium has one less valence electron than silicon. (c) is neither n-type nor p-type, as germanium and silicon have the same number of valence electrons.

QUESTION 8.5 (*Learning Outcome 15*)

(a) Tin is in Group IV (14) and thus has four valence electrons. This is one more than gallium and so there will be electrons in the conduction band. The conductivity will increase.

(b) Zinc has two valence electrons and so will form a p-type semiconductor. There will be fewer electrons in the valence band, but again the conductivity will increase.

QUESTION 8.6 (*Learning Outcome 15*)

(a) The nitrogen atoms will have extra electrons that occupy levels just below the conduction band (Section 8.4) and from where they are easily excited into the conduction band.

(b) To produce yellow, the diamond must be absorbing at the blue end of the visible spectrum.

The role of the nitrogen here is to provide energy levels below the conduction band so that electrons from the valence band can absorb blue light as well as ultraviolet radiation. The band gap is effectively decreased.

Comment: If the nitrogen atoms are trapped as small clusters rather than individual atoms, then they do not form impurity levels as described and the diamonds are colourless.

FURTHER READING

1   D. Johnson (ed.), *Metals and Chemical Change*, The Open University and the Royal Society of Chemistry (2002).

2   L. E. Smart and J. M. Gagan (eds), *The Third Dimension*, The Open University and the Royal Society of Chemistry (2002).

3   M. Mortimer and P. G. Taylor (eds), *Chemical Kinetics and Mechanism*, The Open University and the Royal Society of Chemistry (2002).

ACKNOWLEDGEMENTS

Grateful acknowledgement is made to the following sources for permission to reproduce material in this Book:

Figure 1.2: © NASA; *Figure 3.1*: © Corbis Bettman; *Figures 3.2a, b*: © The Nobel Foundation; *Figure 3.7*: Webster, B. (1990) *Chemical Bonding Theory*, Blackwell Scientific Publications Ltd.; *Figure 6.17*: Atkins, P. W. (1978) *Physical Chemistry*, Oxford University Press/W.H.Freeman & Company; *Figure 8.8*: courtesy of Chris Hartley, IFEC Technology, The Open University; *Figure 8.19*: © Ed Bailey/AP Photo.

Every effort has been made to trace all the copyright owners, but if any has been inadvertently overlooked, the publishers will be pleased to make the necessary arrangements at the first opportunity.

Case *Study*

Molecular Modelling
in Rational Drug Design

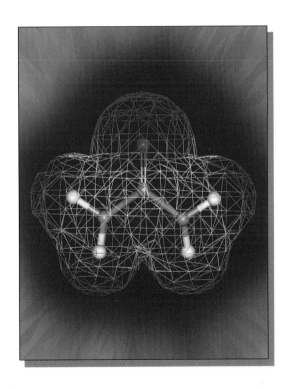

Guy Grant and Elaine Moore

INTRODUCTION

1

Although there are a huge number of drugs available, new drugs are constantly being sought.* Some of these drugs are to treat diseases that have only recently emerged or been recognized, or for which there is no effective medication. Even where medication exists, there are medical reasons for developing alternatives. One reason is that the current treatment may have undesirable side-effects. Older drugs for some mental illnesses, for example, reduced the symptoms of the disease but produced side-effects that some patients were not prepared to tolerate. Another reason is that bacteria and viruses become drug-resistant, perhaps the best-known example being antibiotic-resistant bacteria. Some drugs are effective on some groups of patients but not on others. Finally, researchers may be looking for a drug that is easier or more comfortable to deliver. Many people, for example, dislike injections and would prefer their medicine in the form of tablets.

Alkenes and Aromatics considers the methodology of drug action and design, and shows how the synthesis of one particular drug can be planned.[1]

Until relatively recently, chance played an enormous role in the discovery of medicines. Many drugs were discovered by testing natural products to determine which ones were efficacious. A common starting point was a plant traditionally used to treat a certain condition. Screening of natural products still occurs but nowadays this is to try and broaden the base of possible combinations of functional groups; a natural product may have a particular grouping that had not been thought of when looking for a group that had a particular pharmacological function.

It is more common now, however, to employ structure-based design methods. These start from the identification of a biological pathway in the body or in the bacterium or virus. Many such pathways are regulated by enzymes, proteins that act as catalysts. Recently drugs that interact with DNA have been investigated but the majority of structure-based design has targeted enzymes or receptors, and it is examples of this strategy that we shall consider here.

Enzymes (see Box 1.1) catalyse reactions by binding the reactants so that they can form products via a lower energy pathway than they can when free. Often another molecule, not involved in the reaction, is also bound to the enzyme, and aids the catalytic process. Such a molecule is known as a coenzyme. The reactants and the coenzyme are attached to a particular position in the enzyme known as the active site.

BOX 1.1 Enzymes

Enzymes are catalysts that are found in living systems. They are globular proteins; that is, they are polymers of amino acids folded to form a compact, roughly spherical shape. Within the enzyme are cavities containing active sites to which a substrate attaches and then undergoes chemical change. Enzymes are generally very efficient; nanogram quantities make a considerable difference to the rate of the reaction. They are also specific since the active site will only fit part of a particular molecule or type of molecule.

Some enzymes require assistance from small organic molecules that bind at or close to the active site. These molecules are known as coenzymes; they do not appear in the product and are continually recycled.

Figure 1.1 shows the active site of the enzyme human carbonic anhydrase II. Carbonic anhydrases convert carbon dioxide into hydrogen carbonate ion (HCO_3^-). In the eye this is a critical step in the secretion of aqueous humour (the fluid that fills the eye). Drugs that inhibit this enzyme (e.g. timolol) have been used to treat glaucoma by reducing the secretion.

The active site has been determined by X-ray crystallography and consists of a cone-shaped cavity composed of twisted β-sheets (Figure 1.2) of amino acids (Box 1.2) descending to the zinc atom. In Figure 1.1 the zinc atom is shown as a red sphere.

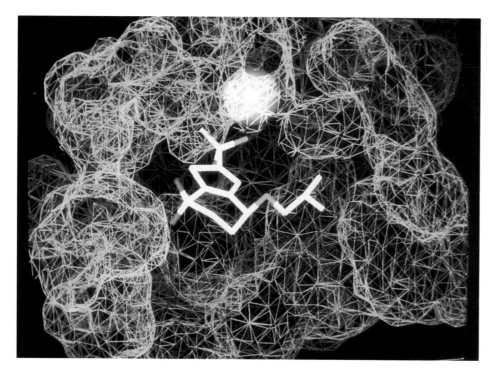

Figure 1.1 Active site of human carbonic anhydrase II with an inhibitor in place. The inhibitor, shown as a separate molecule in the centre, coordinates to the zinc atom via the group shown at the top. It is attached to other atoms in the cavity through non-bonded interactions. Molecules bound to active sites are often bonded via weak interactions such as hydrogen bonds and van der Waals interactions.

BOX 1.2 Secondary structure of proteins

Secondary structures in proteins are regular stable patterns formed by the folding of polypeptide chains. Two common folding patterns found in proteins are illustrated in Figure 1.2. (a) and (b) show, respectively, structural and schematic representations of the α-helix, in which the N—H group of every peptide bond is hydrogen-bonded to the C=O of a peptide bond located four peptide bonds away in the same chain. (c) and (d) show the β-sheet; in this example, adjacent parts of a polypeptide chain run in opposite (antiparallel) directions; they can also run parallel. The different sections of polypeptide chain in a β-sheet are held together by hydrogen bonds between the C=O and N—H groups of peptide bonds in the adjacent sections. (a) and (c) show all the atoms in the polypeptide backbones, but the amino acid side chain groups are denoted simply by R. In contrast, (b) and (d) show the α-helix and β-sheet as ribbons.

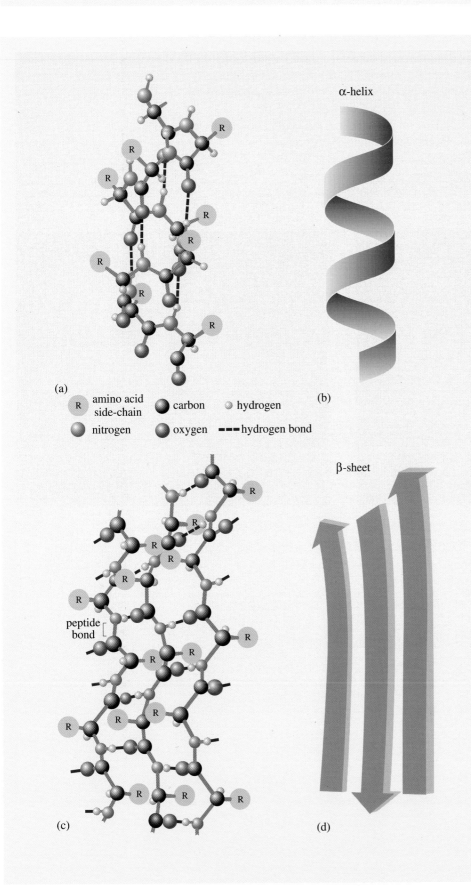

α-helix

β-sheet

(a)

R — amino acid side-chain   carbon   hydrogen

nitrogen   oxygen   --- hydrogen bond

(b)

peptide bond

(c)

(d)

Figure 1.2
An α-helix (a) and (b) and a
β-sheet (c) and (d) of amino
acids. In the β-sheet, two or
more polypeptide chains run
alongside each other and are
linked in a regular manner by
hydrogen bonds.

The blue net in Figure 1.1 is formed from the van der Waals radii of the atoms lining the cavity. The van der Waals radius of an atom is obtained from the volume of the atom when it is in non-bonded contact with other atoms. Non-bonded atoms cannot approach closer than their combined van der Waals radii, so the surface indicated by the net gives a boundary of closest approach for the van der Waals surface of incoming molecules. A van der Waals surface for sialic acid, a molecule important in the development of the anti-flu drug Relenza® (Section 3), as produced by WebLab ViewerLite, is shown in Figure 1.3. In Figure 1.3a, the van der Waals sphere around each atom is in that atom's colour, showing how the net is made up from individual atoms' contributions. Figure 1.3b shows the whole surface as it is usually represented in one colour.

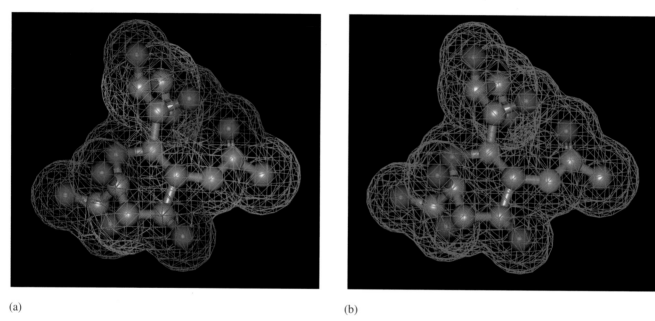

(a)   (b)

Figure 1.3 Sialic acid showing van der Waals surface (a) with contributions from each atom shown in the atom colour and (b) in blue. 🖳

Drug design is easier if the structure of the active site is known but there are important enzymes whose structure is unknown and we start this Case Study by looking at methods for designing drugs to target unknown active sites.

sialic acid

STRUCTURE-BASED METHODS FOR UNKNOWN DRUG ACTIVE SITES

2

2.1 Quantitative structure–activity relationships (QSAR)

This method is one of the earliest approaches to structure-based design. It relies on finding a correlation between the physical properties of a compound and its activity as a drug. A typical physical property used is the *partition coefficient* between octanol and water — that is, the relative concentrations of a compound in the two solvents when it is allowed to distribute itself between them. This is used as a crude measure of the preference of the compound to dissolve in lipids (fats) in the body. This solubility in lipids is important as it can determine where in the body the drug can reach.

The activity towards a target enzyme for a large number of compounds is measured by obtaining the equilibrium constant for the reaction:

$$\text{molecules} + \text{enzyme} \rightleftharpoons \text{molecule–enzyme complex} \tag{2.1}$$

Values of various other physical properties for each of the compounds are then either measured or obtained from the scientific literature or databases of properties. An equation is derived which expresses the activity as in Equation 2.1 or the biological activity as manifested in physiological reactions in terms of the physical properties. The physical properties used and the weight given to them are adjusted until the best fit to all the compounds is obtained.

As an example of this method, we take a study of the drug clonidine (Structure **2.1**), primarily used in the treatment of high blood pressure.

Clonidine reduces blood pressure by acting on the central nervous system, but it also contracts blood vessels in the peripheral nervous system, and this leads to an increase in blood pressure. The biological activity of a series of related compounds on the central nervous system was found to correlate with the partition coefficient between octanol and water and the biological activity on the peripheral nervous system. The study indicated that there was an optimum value for the partition coefficient, and that central and peripheral nervous system activity could not be separated, suggesting that to find a drug that only acts on the central nervous system it would be necessary to try compounds unrelated to clonidine. Clonidine attaches to the same type of receptor in both the central and peripheral nervous systems — hence the correlation between central and peripheral nervous system biological activity. To attach to the receptor in the central nervous system, however, it has to cross the blood–brain barrier and thus must dissolve in lipids to reach its target. This leads to the correlation of central nervous system activity with lipid solubility.

QSAR used bulk properties of the potential drug molecules. A more recent method, 3D-QSAR, correlates activity with properties that depend on particular parts of the molecular structure such as the electrostatic potential. In 3D-QSAR a set of molecules of known activity is superimposed by computer so that similar groups in each molecule are in the same place. A box is drawn that contains all the molecules

clonidine
2.1

and leaves a few hundred pm beyond the atoms at the extremes. The box is divided up into a lattice of n points along each side with the points typically 200 pm apart. A box containing one molecule is shown in Figure 2.1.

Many properties are calculated by introducing a theoretical probe molecule or group of atoms at each point in the lattice, and determining the energy of the interaction between the probe and the molecules under study. For example, an ion might be used to determine the electrostatic interaction, or a neutral atom used to determine the intermolecular forces. Such calculations rely on molecular mechanics methods and can be performed extremely rapidly, so that the interaction of every molecule with a probe at each grid point can be ascertained. Suppose a cubic grid of $n \times n \times n$ points were used and the interaction with two probes at every point calculated. There would now be $2n^3$ variables to fit to the activity. Inevitably, however, the properties will only correlate strongly at some points and the rest of the points can effectively be ignored. An equation relating the properties at the significant points to the activity can then be obtained and used to predict the activity of other molecules.

As well as properties derived from molecular mechanics on probe molecules, properties obtained from molecular orbital calculations have also been used, for example the partial charge on a particular atom. In a study of the power of aromatic molecules containing nitro and amino groups to cause mutations, the researchers considered the contributions of atomic orbitals on particular atoms to molecular orbitals of the aromatic molecules. One of the factors found to be important was the amount of an atomic orbital on the atom occupying site 1 (S1) in Figure 2.2 that contributed to the lowest unoccupied molecular orbital.

Figure 2.1
Molecule in a typical 3D-QSAR lattice.

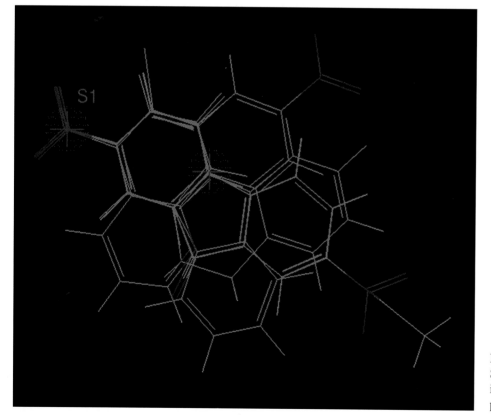

Figure 2.2
Several superimposed molecules used in a 3D-QSAR study of mutagenic potency, showing the site S1.

2.2 Active analogues

As with 3D-QSAR, this method starts from a collection of active molecules. Instead of simply looking at the properties, the method starts by obtaining common structural features for the set of molecules. A common set of functional groups in a particular geometric arrangement is known as a **pharmacaphore**. Pharmacaphores have been used by several groups studying dopamine transporter proteins. Dopamine is a neurotransmitter of importance in Parkinson's disease, Alzheimer's disease and some types of drug abuse. Because the dopamine transporter proteins are embedded in cell membranes, it is very difficult to obtain their structure as they cannot be easily crystallized.

In a study of possible drugs to treat cocaine abuse, the pharmacaphore shown in Figure 2.3 was obtained.

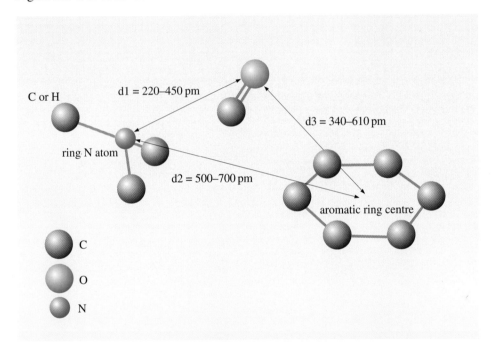

Figure 2.3
A pharmacaphore model derived from cocaine and its analogues.

Previous structure–activity studies had shown that a benzene ring, an ester or small hydrophobic group, and a nitrogen or oxygen atom, were important features. The distances between these groups in Figure 2.3 were determined by comparing the distances in low-energy conformations of cocaine (Structure **2.2**) and a cocaine antagonist (Structure **2.3**) (see Box 2.1).

cocaine
2.2

a cocaine antagonist
2.3

BOX 2.1 Agonists and antagonists

An agonist binds to a receptor molecule and produces the same response as the drug or naturally occurring compound which the receptor is designed to fit.

An antagonist binds to a receptor and blocks the response.

Thus a cocaine antagonist will bind to the same site as cocaine and reduce the effect of the cocaine.

A pharmacaphore gives some insight into the nature of the active site if we assume that all the active molecules bind at the same site. The accessible volume of the binding site can be estimated by taking the minimum volume required to enclose the complete set of superimposed active molecules. The binding site is presumed also to contain groups that interact with the elements of the pharmacaphore in a geometrical relationship that will match the pharmacaphore.

Further information can be obtained by looking at molecules apparently similar to the active molecules but which only bind poorly to the receptor site, and noting differences in size and functional groups between the active and inactive set.

METHODS FOR KNOWN ACTIVE SITES

3

Increasing numbers of active sites have now had their structures determined. In such cases designing molecules becomes easier, and indeed much information can be obtained simply by viewing how a molecule fits into the active site.

3.1 GRID

The GRID method, like 3D-QSAR, uses a grid of points at which probe molecules are placed. In this case, however, it is the interaction with the active site of the enzyme rather than with potential drug molecules that is calculated. The most spectacular use of the GRID method was in the discovery of the anti-influenza drug Relenza®.

'Flu' viruses are spherical in shape with a core of RNA within a shell of proteins and lipids. The whole virus is covered by hundreds of spikes, and is shown schematically in Figure 3.1.

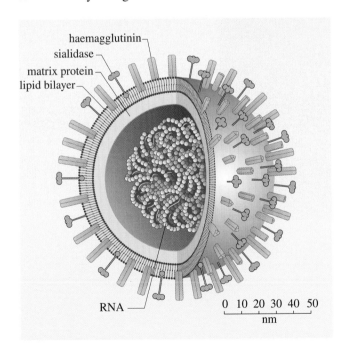

Figure 3.1
Schematic diagram of an influenza virus.

The spikes are of two types. One type (haemagglutinin) has a vital role in the first stages of infection, the penetration of host cells. The second type of spike is made of an enzyme called sialidase, whose role is to help newly formed viruses escape from the host cell so that they can move on to infect other cells. As part of this process, sialidase cleaves off sialic acid from oligosaccharides (short chains of carbohydrate units). These sialidase spikes have clefts that contain an active site, into which bound sialic acid (Structure **3.1**) fits during the cleavage process. This active site

sialic acid
3.1

135

seems to remain constant across different influenza strains, the amino acid residues surrounding the site providing the variation.

The researchers set out to find a compound that would inhibit the action of sialidase, thus preventing infection from spreading from the initial host cell. They were therefore looking for a compound that would bind to the active site, preventing sialic acid from reaching it and so preventing the cleavage of sialic acid from oligosaccharides. The starting point was the structure of sialidase with sialic acid bound to its active site. The structure of sialidase itself was first obtained; then a crystal was immersed in a solution of sialic acid and the structure determined again. The active site cavity with the sialic acid in place was examined carefully, and it was found that there was space in the cavity where an extra group could be added to sialic acid. The active site is shown in Figure 3.2.

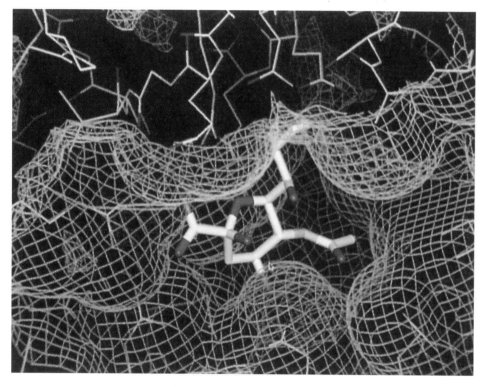

Figure 3.2
Structure of the sialidase active site with sialic acid bound, showing the region of the pocket with space for fitting an extra group to the right of the molecule. The blue net is a van der Waals surface and indicates the accessible surface of the cavity.

The GRID program was then tested by using a carboxylic acid probe to explore the cavity. For this probe as well as electrostatic and intermolecular interactions we can consider hydrogen-bonding of the group to the active site through, for example, the C=O of the carboxylic acid group and a suitable hydrogen in the cavity. The H group of the —COOH can also be lost to an atom in the active site. A site where interaction with the probe was strong was identified exactly where the —COOH group of sialic acid sits. GRID was then used again with an —NH$_2$ probe. Again this probe can hydrogen-bond to the active site but can also gain a hydrogen from the site to form —NH$_3^+$. This probe found several sites where interaction was high. As a result of this investigation, the researchers decided to adapt a compound called Neu5Ac2en (Structure **3.2**), which had been shown to be a weak inhibitor of sialidase but did not work in living systems. The —OH group attached to the ring in Neu5Ac2en was replaced by —NH$_2$ or by the guanidino group (Structure **3.3**).

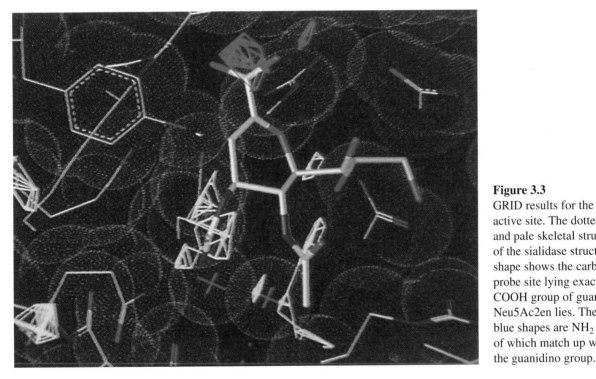

Neu5Ac2en
3.2

guanidino group
3.3

The guanidino group could occupy two of the sites identified by the —NH₂ probe, and so would potentially attach more strongly and be a better inhibitor. The GRID results showing the carboxylic acid site and several —NH₂ sites are shown in Figure 3.3. The carboxylic acid probe site exactly matches the position of the —COOH group of guanidino–Neu5Ac2en and two of the —NH₂ probe sites shown match up with the ends of the guanidino group.

Figure 3.3
GRID results for the sialic acid active site. The dotted blue spheres and pale skeletal structures are parts of the sialidase structure. The red shape shows the carboxylic acid probe site lying exactly where the COOH group of guanidino–Neu5Ac2en lies. The various light blue shapes are NH₂ probe sites, two of which match up with the ends of the guanidino group.

Guanidino–Neu5Ac2en proved not only to be a more effective inhibitor than Neu5Ac2en but was also found to work when used on patients, and was eventually marketed as Relenza®.

3.2 Multicopy simultaneous search

Like GRID, this method uses probe molecules to explore the active site of the enzyme. Instead of calculating the energy at points on a grid, however, the probe is allowed to move around until it finds a minimum energy position. The procedure can be used to minimize the positions of hundreds of probe molecules at once because each probe is only allowed to interact with the enzyme and not with other probes. This method will give similar results to the GRID method; the low-energy sites corresponding to the most favourable grid points. In principle, it has a potential advantage over GRID in that it is possible to allow part of the active site to move during minimization. This would mimic the experimental situation more closely but is expensive of computer resource. Figure 3.4 shows the result of a typical multicopy simultaneous search on a protein with molecules clustering where interaction with the protein is greatest.

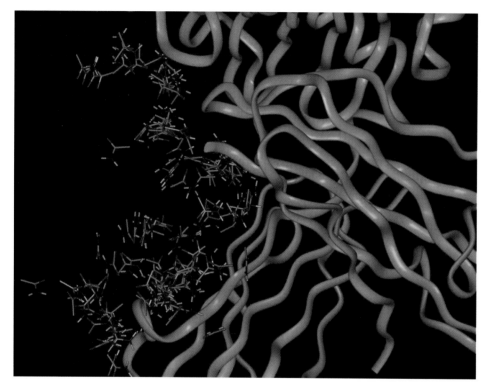

Figure 3.4
Result of a multicopy simultaneous search on a human glycoprotein. The purple ribbon represents the protein. The probe molecules are shown as stick models.

A recent (2000) study used this method to identify potential inhibitors of HIV-1 integrase. Current HIV drugs target enzymes known as reverse transcriptase and protease as you will see in Section 4. HIV-1 integrase inserts viral DNA into the host chromosome, and inhibitors of this enzyme would provide an alternative class of drugs. In the reported study, the active site of the enzyme was allowed to adopt different conformations in response to probe molecules. Several possible inhibitors were identified.

EXAMPLES OF STRUCTURE-BASED DESIGN

4

In this Section we follow the development of two drugs, one for AIDS and one for cancer. The stories behind these illustrate the interdisciplinary nature of drug research, with physicians, biologists, biochemists, X-ray crystallographers and chemists all playing a part. The initial goal of such research is often a **lead compound** (pronounced 'leed', i.e. a starting compound, *not* one containing lead, Pb). The lead compound has some activity but is fairly simple and has several positions on its carbon framework where functional groups can be modified. An iterative process ensues, during which modifications are made and the molecule is then tested for its activity in binding to an active site or receptor, and for its biological activity. This process is continued until a compound of promising high biological activity is found. Further cycles of the process may be needed if the compound is insoluble or too soluble or insufficiently discriminating. Eventually a potential drug is sent for clinical testing, but even at this stage it may fail because of unwanted side-effects or because it is metabolized before it reaches its target, necessitating a return to the design stage.

Our first example is the development of HIV-protease inhibitor drugs.

4.1 HIV-protease inhibitor drugs

HIV-protease inhibitors are one class of drugs used to treat AIDS, but the development of these drugs starts with a search for a drug for a different condition — high blood pressure.

4.1.1 Renin inhibitors

Renin is an enzyme involved in the production of angiotensin, a hormone found in the kidney that raises blood pressure. Renin works in concert with another enzyme, angiotensin-converting enzyme (ACE). ACE-inhibitors have been used successfully to treat high blood pressure since 1979, and researchers were exploring the possibility of renin-inhibitors as an alternative. Renin works on a tripeptide sequence (a series of three amino acids linked together) in a protein-precursor to angiotensin. The approach used in designing an inhibitor was to find a similar tripeptide sequence but with the peptide linkage (Structure **4.1**) that was normally cleaved by renin replaced by a group that would not be cleaved, so that the molecule remained bound to the enzyme. Highly effective inhibitors were found but unfortunately in clinical trials it was discovered that the compounds were metabolized by the liver and never reached the kidneys where the renin is located. As you will see later, however, this research was of use in the search for anti-HIV drugs.

4.1.2 Approaches to anti-HIV drugs

Researchers have set out to target almost all stages in the life cycle of HIV, but the drugs currently available fall into two main types. The earliest developed were the reverse transcriptase inhibitors, e.g. AZT (Structure **4.2**), which stop the replication of viral genes. Transcription is the process in which the DNA base sequence is

peptide linkage
4.1

AZT
4.2

copied to form RNA (see Box 4.1). Viruses such as HIV contain RNA rather than DNA as their genetic material. AZT and similar drugs such as Viread™, recently approved for clinical use in the UK (2002), are based on the building blocks of DNA/RNA, a base plus a sugar. There is also a group of drugs based on different structures which also inhibit reverse transcriptase, e.g. nevirapine.

BOX 4.1 DNA and RNA

DNA has a double helix structure, each strand of which is a polymer. The monomer units of these polymers consist of a phosphate group, a sugar molecule (deoxyribose) and a base (adenine, guanine, cytosine or thymine). The two strands are linked together by hydrogen bonds between the bases. The sequence of base pairs in a DNA molecule which go to make up a gene has a direct relationship with the amino acid sequence of the polypeptide specified by the gene. The production of a polypeptide from the genetic information in DNA proceeds via RNA. RNA is a polymer similar to DNA but with a different sugar (ribose) and with uracil instead of thymine.

Genetic information is copied from DNA to RNA in a process called *transcription*.

The other type of anti-HIV drugs are the HIV-protease inhibitors such as saquinavir, indinavir, nelfinavir, ritonavir and amprenavir.

HIV-1 protease is a virally coded enzyme that cleaves polypeptides (chains of amino acids) to make them ready to form proteins during virus assembly and maturation. In 1988, gene technology was used to create a mutant HIV protease. The mutant enzyme resulted in the virus replication system only producing immature viral cells incapable of infecting human cells. Researchers set out to find a drug that would inhibit HIV protease, thus preventing HIV particles from maturing. Figure 4.1 shows an HIV particle in schematic form. HIV-protease inhibitors interrupt the formation of the virus-specific proteins shown on the surface.

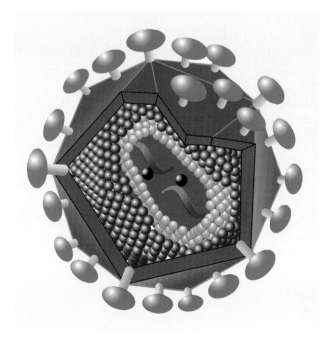

Figure 4.1
Artist's impression of an HIV particle, depicting RNA (orange ribbons) and reverse transcriptase (grey sphere) in the centre and virus-specific proteins on the surface.

The crystal structure of HIV-1 protease was solved in 1989 and was shown to be similar to those of a family of enzymes known as aspartic proteases, of which renin is a member. Aspartic proteases have an active site containing two aspartic acid residues.

4.1.3 Designing an HIV-1 protease inhibitor

Because of the similarity of the active sites of renin and HIV protease, the approach used was one that had been developed for renin inhibitors. HIV protease cleaves the linkage between two amino acids — phenylalanine (Structure **4.3**) and proline (Structure **4.4**).

The initial strategy was to find a molecule that resembled phenylalanine so that it would bind to the active site but whose differences in structure meant that it would not be cleaved. The X-ray structure indicated that there was a C_2 axis in the active site, and so a molecule with a C_2 axis, or which was close to C_2 symmetry, was sought. The initial structure chosen was Structure **4.5**, in which each side of the molecule resembles phenylalanine.

This structure was modified by replacing one of the hydrogens on each of the NH_2 groups. The most potent inhibitor in this series had both hydrogens replaced by cbz—Val (Structure **4.6**), the amino acid valine protected by a protecting group whose trivial name is carbobenzoxy.

phenylalanine
4.3

proline
4.4

4.5

cbz–Val
4.6

This compound was co-crystallized with the protease and the crystal structure determined. The structure confirmed the symmetry-based mode of binding for diamino alcohols, but also showed that the spacing between the two NH groups was too short to make good hydrogen bonds with the enzyme. The hydrogen-bond interactions are shown in Figure 4.2 (overleaf).

To increase the distance between the NH groups, the central —CHOH group of Structure **4.5** was replaced by a diol grouping, —CHOH—CHOH—. Inhibitors based on the new skeleton (Structure **4.7**) were more powerful but their usefulness was limited by their poor solubility. The solvent-accessible surface of HIV protease with an inhibitor was examined (Figure 4.3 overleaf). This is done by positioning a solvent molecule so that its van der Waals surface is just in contact with the van der Waals surface of the active site. The solvent molecule is then allowed to move, keeping the two van der Waals surfaces just in contact. The path of the solvent molecule thus plots out a surface representing the closest approach of the solvent to the active site, known as the solvent-accessible surface.

This exercise indicated that the ends of the inhibitor were exposed, and that the solubility could be enhanced by modifying the end groups. One compound produced by this route entered clinical trials as an intravenous treatment. This compound had the basic diamino, diol structure but the X groups were as shown in Structure **4.8**.

4.7

4.8

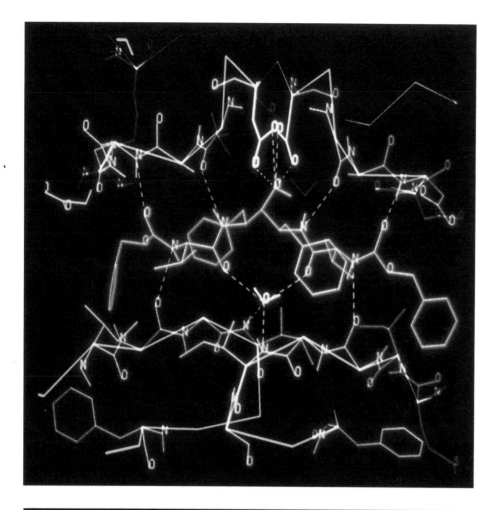

Figure 4.2
H-bonding interactions between an inhibitor and HIV protease. The active site carboxylate groups (white) interact with the central OH group on the inhibitor. The inhibitor is the molecule in pink stretching across the centre of this figure.

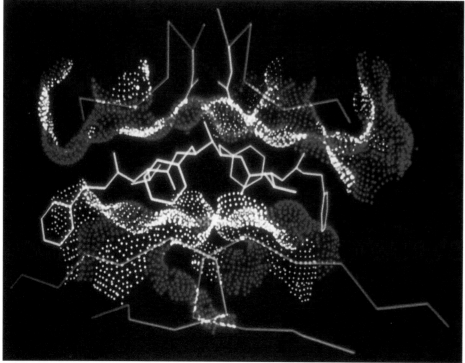

Figure 4.3
View of a 1 000–1 500 pm thick section of the active site of the HIV-1 protease. Red dots represent the hydrophilic solvent-accessible surface and blue dots the hydrophobic solvent-accessible surface obtained from a model with the inhibitor removed. An inhibitor is shown in the site.

Placing this compound in the active site and using modelling methods to determine how they best fitted led to the discovery that the OH groups could hydrogen-bond to carboxylate ($-COO^-$) groups in the active site in two ways. Each $-OH$ could bind to one $-COO^-$ or one $-OH$ could bind to the two available $-COO^-$ groups and the other OH play only a minor role in bonding.

Replacing one $-OH$ group in Structure **4.7** by H led to a better inhibitor. Several modifications were then made on the X groups, and eventually a compound, Structure **4.9**, was entered into clinical trials as a treatment that could be taken orally.

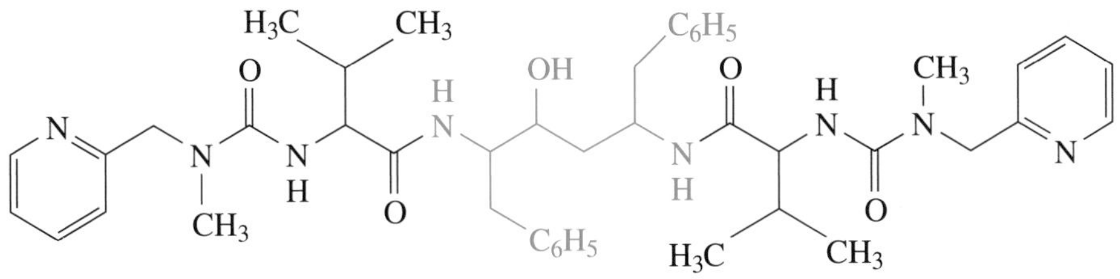

4.9

In Structure **4.9** we have picked out the basic skeleton from Structure **4.7** in green. Further modifications finally led to the drug ritonavir. Ritonavir was approved by the USA Food and Drug Administration (FDA) in March 1996. In ritonavir, Structure **4.10**, you can see the central group discussed in this example, with modified end groups.

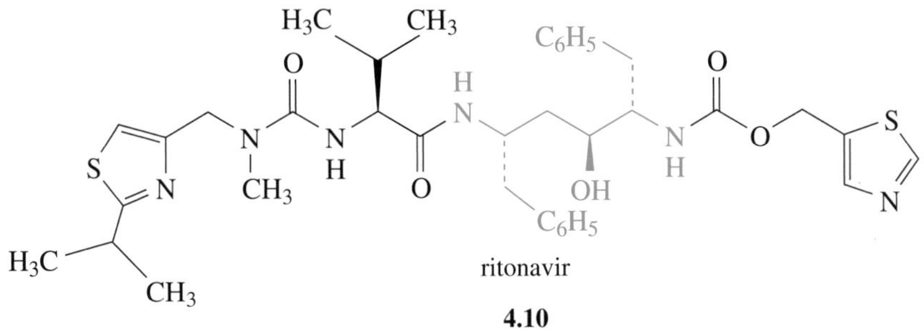

ritonavir

4.10

The first HIV-protease inhibitor drug saquinavir, was approved by the USA FDA in December 1995. The development of this drug started from an initial tripeptide lead compound. The structure (**4.11**) is given below:

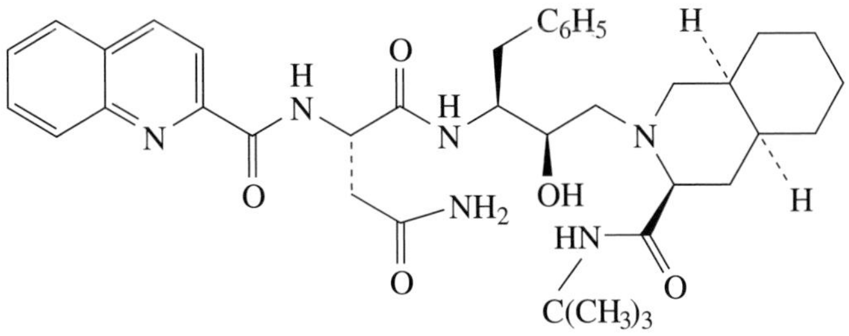

saquinavir

4.11

HIV is a virus that rapidly becomes drug-resistant; it took only 18 months for the virus to develop resistance to AZT. The current treatment, therefore, often involves a cocktail of several drugs acting in different ways, that is a combination of protease inhibitors and reverse transcriptase inhibitors. At the same time research continues into possible new drugs, for example the HIV-1 integrase inhibitors mentioned in Section 3.

4.2 De novo drug design

Our second example illustrates a strategy known as *de novo* drug design in which a lead compound is designed by fitting a carbon backbone and functional groups into the active site of an enzyme. The work we describe aimed to find a drug that would inhibit an enzyme called thymidylate synthase (Figure 4.4). Inhibitors of thymidylate synthase have been shown to be antitumour agents. This enzyme catalyses the rate-determining step in the conversion of deoxyuridylate monophosphate (dUMP) (Structure **4.12**) into thymidylate monophosphate. dUMP is the base uracil found in RNA, attached to a deoxyribose molecule joined to a phosphate group. In thymidylate monophosphate (Structure **4.13**), the uracil has been converted into thymine, the corresponding base found in DNA. The enzyme binds both dUMP and a coenzyme 5,10-methylenetetrahydrofolate (Structure **4.14**).

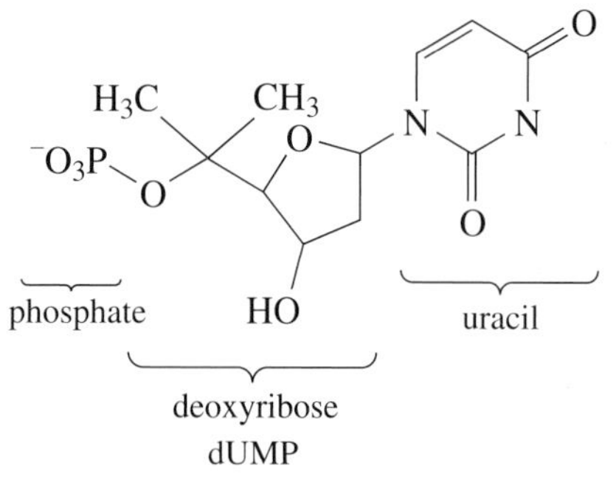

dUMP

4.12

thymidylate monophosphate

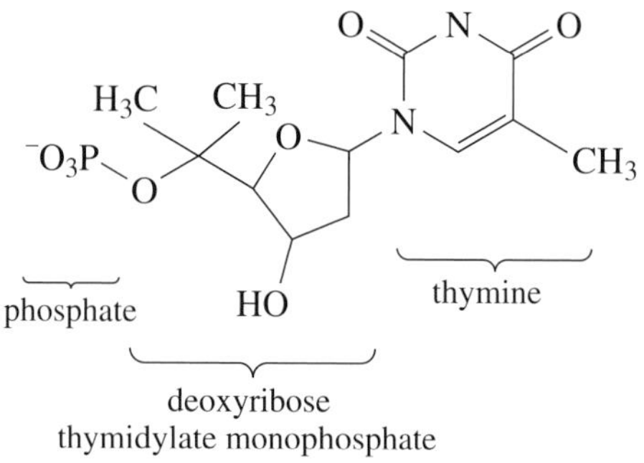

4.13

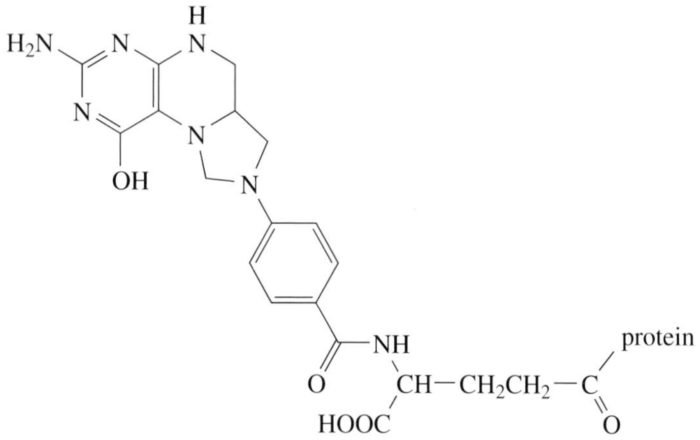

5,10-methylenetetrahydrofolate

4.14

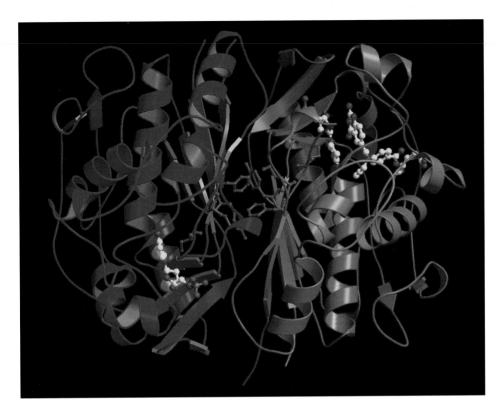

Figure 4.4
Thymidylate synthase. Structures such as α-helices and β-sheets are shown in a ribbon representation.

The researchers aimed to inhibit the enzyme by blocking the coenzyme site.

They started from the X-ray structure of thymidylate synthase (from *E. coli* bacteria) complexed with an inhibitor. A methyl probe was used with the GRID program to produce a contour map of strong interaction with the active site. On the computer screen, a naphthalene molecule (Structure **4.15**) was bound into the active site, so that it had maximum overlap with the contours.

At one position on the naphthalene rings there was a nearby group on the enzyme that could hydrogen-bond. An NH group was placed at this position. On another position a C=O group was added to hydrogen-bond to a bound water molecule in the active site. These two groups were linked to give Structure **4.16**.

Empty space was identified adjacent to one of the rings and filled with a nitrogen atom attached to two alkyl groups. A nitrogen atom was chosen rather than a carbon atom because it would not create a chiral centre when substituted, thus simplifying the synthetic chemistry required. Figure 4.5 (overleaf) shows the fit of this fragment on the methyl probe contours, shown as the pink grid.

Space still remained and so one of the *N*-alkyl groups was made to be benzyl, —CH₂Ph. To complete the molecule, a solubilizing group (Structure **4.17**) was added, which modelling indicated would sit at the interface of the protein and bulk solvent.

naphthalene
4.15

4.16

4.17

This molecule formed the lead compound Structure **4.18**.

4.18

This molecule was synthesized and tested for activity. The crystal structure of the complex of this compound with the enzyme was then determined. The compound was found to bind as expected with one major exception. The $-C=O$ group interacts unfavourably with a $-C=O$ group in the active site and breaks a hydrogen bond between the enzyme $-C=O$ group and the bound water molecule.

The next stage then was to replace the $-C=O$ group with a nitrogen atom. This new structure bound to the active site as predicted, and was a better inhibitor. It was therefore used as a guide, and a number of derivatives were designed that were intended to fill the remaining space in the deep part of the active site. Figure 4.6 shows the crystal structure of the new lead compound complexed with the enzyme; the important hydrogen bonds are indicated as white lines.

This lead compound was further modified to produce several new compounds. One of these, Structure **4.19**, was sent for clinical trials as an antitumour agent.

4.19

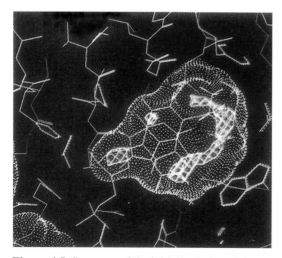

Figure 4.5 Structure of the initially designed ring (green) overlaid on GRID contours in the active site of thymidylate synthase. The dotted surface represents the active site. The orange stick structures are parts of the enzyme.

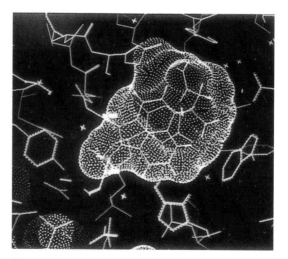

Figure 4.6 Crystal structure of lead compound complexed with thymidylate synthase.

THE FUTURE

5

Research of the type illustrated here will become more common. One project that will have an immediate effect is the Human Genome Project. A genome is a cell's total content of genetic material. One product of the Human Genome Project is insight into the structure of the membrane-bound proteins whose structures are difficult to determine. In a related effort, attempts to associate a biological function with all products produced by genes will increase our understanding of the biochemical pathways associated with a particular disease. This should lead to identification of new targets.

Along with the Human Genome Project, the genomes of a variety of other organisms are being studied. Information on the genomes of bacteria can provide insight into the metabolic pathways of these organisms and thus identify different strategies for drug design.

Many drugs work better for one subset of the population than for another. There has been speculation that in the future, you could be offered drugs tailored to your particular genetic make-up.

We have concentrated on drugs that target enzymes, but already there are drugs that target DNA instead, and this approach could grow in the future.

Finally, the future may bring not only new drugs but new methods of delivery, e.g. through skin patches or nasal sprays rather than pills or injections, and drugs that are activated only when they reach the part of the body where they are needed. Methods in development include bioerodable implants, for example a thin wafer impregnated with an anticancer drug that is surgically placed at the site of a brain tumour and slowly eroded by brain fluids, releasing the drug as it erodes. An insulin-laden plastic implant that is activated by ultrasound or magnetic fields has been suggested. Another interesting drug delivery vehicle is a liposome, a microscopic bubble of lipids surrounding an aqueous interior containing the drug which could be directed to a particular site.

FURTHER READING

1 P. G. Taylor and J. M. Gagan (eds), *Alkenes and Aromatics*, The Open University and the Royal Society of Chemistry (2002).

ACKNOWLEDGEMENTS

Grateful acknowledgement is made to the following sources for permission to reproduce material in this Case Study:

Figure 1.1: Reprinted with permission from *Journal of Medicinal Chemistry*, Vol. 37, No. 8, p. 1047, 1994, fig. 11, American Chemical Society; *Figure 2.1*: Reprinted from Kubinyi, H., 'QSAR and 3D-QSAR in drug design, part 1: methodology', *Drug Discovery Today*, Vol. 2, No. 11, 1997, p. 457, Copyright © 1997 with permission from Elsevier Science; *Figure 2.2*: © Accelrys, Inc.; *Figure 2.3*: Reprinted with permission from *Journal of Medicinal Chemistry*, Vol. 43, No. 3, Copyright © 2000 American Chemical Society; *Figure 3.1*: Kaplan, M. M. and Webster, R. G. 'The epidemiology of influenza', *Scientific American*, Vol. 237, No. 6, December 1977, Scientific American, Inc., © Bunji Tagawa, used by kind permission of the Estate of Bunji Tagawa; *Figure 4.1*: 'Halting viral replication', *Beyond Discovery*, National Academy of Sciences, Copyright © 2000 National Academy of Sciences, http://www.beyonddiscovery.org/beyond/beyonddiscovery.nsf; *Figure 4.2*: Reprinted with permission from *Journal of Medicinal Chemistry*, Vol. 37, No. 8, p. 1041, 1994, fig. 6, American Chemical Society; *Figure 4.3*: Reprinted with permission from *Journal of Medicinal Chemistry*, Vol. 37, No. 8, p. 1042, 1994, fig. 7, American Chemical Society; *Figure 4.4*: Courtesy of Amy C. Anderson; *Figure 4.5*: Reprinted with permission from *Journal of Medicinal Chemistry*, Vol. 37, No. 8, p. 1051, 1994, fig. 16, American Chemical Society; *Figure 4.6*: Reprinted with permission from *Journal of Medicinal Chemistry*, Vol. 37, No. 8, p. 1051, 1994, fig.18, American Chemical Society.

Every effort has been made to trace all the copyright owners, but if any has been inadvertently overlooked, the publishers will be pleased to make the necessary arrangements at the first opportunity.

INDEX

Note Principal references are given in bold type; picture and table references are shown in italics.